Astronomy and Astrophysics for the 1980's

VOLUME 1:
Report of the Astronomy Survey Committee

Astronomy Survey Committee

Assembly of Mathematical
and Physical Sciences

National Research Council

NATIONAL ACADEMY PRESS
Washington, D.C. 1982

Front Cover   *False-color map of radio emission from the galaxy 3C449 recorded by the Very Large Array (VLA) radio telescope of the National Radio Astronomy Observatory near Socorro, New Mexico. Colors are correlated with the intensity of radiation at a wavelength of 20 cm; the most intensely emitting regions are shown in red. The map reveals highly collimated jets of matter connecting an unresolved galactic nucleus to outlying "lobes" of ejected gas. (Photo courtesy of the National Radio Astronomy Observatory)*

Back Cover   *False-color VLA map of 20-cm radio emission from the supernova remnant Cas A, tracing the remains of a cataclysmic stellar explosion that occurred some 350 years ago in our Galaxy in the direction of the constellation Cassiopeia. The radiation is generated in shock waves sent through the surrounding interstellar gas by the expanding shell of stellar debris. (Photo courtesy of the National Radio Astronomy Observatory)*

Frontispiece   *Five of the 27 antennas, each 25 m across, that comprise the Very Large Array (VLA) in Socorro, New Mexico. Construction of the VLA, now operated by the National Radio Astronomy Observatory, was the highest-priority recommendation of the Greenstein report. (Photo courtesy of the National Radio Astronomy Observatory)*

Library of Congress Cataloging in Publication Data

National Research Council (U.S.) Astronomy Survey Committee.
  Astronomy and astrophysics for the 1980's.

  Includes index.
  Contents: v. 1. Report of the Astronomy Survey Committee.
  1. Astronomy.   2. Astrophysics.   I. Title.
  QB43.2N.38 1982   520   82–8014
  ISBN 0–309–03249–0 (v. 1)   AACR2

Available from:

NATIONAL ACADEMY PRESS
2101 Constitution Ave., N.W.
Washington, D.C. 20418

Printed in the United States of America

January 1982

Dr. Frank Press
President
National Academy of Sciences
Washington, D.C. 20418

Dear Dr. Press:

I take pleasure in transmitting to you the report of the Astronomy Survey Committee, *Astronomy and Astrophysics for the 1980's.* This report contains the recommendations resulting from the third decennial study of astronomy and astrophysics to be carried out by the National Academy of Sciences.

The Whitford report, *Ground-Based Astronomy: A Ten Year-Program* (1964), presented the first Academy survey of the field and recommended a program for the construction of facilities for optical and radio astronomy, calling attention particularly to the need for increased numbers of optical telescopes of intermediate size. The Greenstein report, *Astronomy and Astrophysics for the 1970's* (1972), treated both space and ground-based programs and for the first time established priorities ranging across all the fields of astronomy and astrophysics. Its recommendations led to the construction of the Very Large Array (VLA) radio telescope near Socorro, New Mexico; the launch of three High Energy Astronomical Observatory (HEAO) satellites for observations of x rays, gamma rays, and cosmic rays from space; the improvement of detectors for optical and infrared radiation; and a number of other advances. The present survey was requested by the National Science Foundation and the National Aeronautics and Space Administration, which shared its funding.

The Assembly of Mathematical and Physical Sciences accepts this report as the consensus of the U.S. astronomical community and believes that it will be ranked quickly with the Whitford and Greenstein reports as an indispensable blueprint for the future of astronomy and astrophysics during their respective decades. The recommendations deserve the prompt attention and serious response of the agencies of our government. Unfortunately, the immediate future of support for astronomical science is clouded by the present austerity in budgeting for nearly all federal programs; one may hope, however, that the purposes of the present fiscal policies will be achieved in a reasonably short period and that a healthier base of federal scientific support will then be restored. Looking ahead to that time, the Assembly discerns two broader concerns embodied in the report that deserve particular emphasis.

The first relates to the future balance of funding provided to astronomy and astrophysics by the two federal agencies primarily charged with its support. The National Aeronautics and Space Administration was able, during the 1970's, to compile an impressive record of achievement in space astronomy and astrophysics despite severe budgetary constraints and the consequent inability to undertake numbers of outstanding proposed projects. Although some have been temporarily deferred, many of the facilities that will form the cornerstone of research in the coming decade (such as Space Telescope) are a testimony to the commitment of the federal government in support of space science during the 1970's at the level required for steady advance. As is made clear in the report, sustaining such a level of support is necessary to preserve the vigor of U.S. astronomy during the 1980's.

By contrast, the annual budget of the Astronomy Division of the National Science Foundation (NSF) has for some years been less successful in providing the support necessary to sustain the vigor of programs in ground-based astronomy. The decade of the 1970's saw the initiation of only one major NSF construction project in astronomy, the VLA. This magnificent instrument, the first recommendation of the Greenstein report, has already begun to make dramatic contributions to research. However, funding for the construction of a highly capable millimeter-wave radio telescope, another prominent recommendation of the Greenstein report, has not yet received approval, although such authorization was widely expected to appear in the 1982 federal budget. Moreover, the NSF Astronomy Division operations budget—burdened by having to absorb the expenses of the Sacramento Peak Observatory during the 1970's and those of VLA operations from now on—has declined in real dollars to a level that threatens the productive operation both of the National Astronomy Centers and of a grants program vital to basic astronomical research at U.S. universities. There is a serious risk that the unique capabilities for astronomical research established in the United States with the backing of Congress and the Executive Branches of government will be dissipated. Commitment to re-

vi

store the health of U.S. programs in ground-based astronomy is urged in this report.

The second concern we see reflected in this report, and one that extends throughout the scientific community, relates to the need for expanding national support of new-technology developments. Most of the exciting discoveries in astronomy during recent decades are the direct result of advances in technology. In some cases, these advances resulted from programs directed specifically toward astronomical research; in others, they resulted from the adaptation of new technology originally developed for other scientific, industrial, or military purposes. Thus, a broadly based program of technological development is no less important than the specific programs recommended in this report, if astronomical research is to move forward in the decade ahead. In fact, we believe that this is true of science in general; our national capability in science depends on a renewed commitment to a broadly based program of technological development. In a time of general retrenchment in the initiation of new starts, investment in development of new technology and in design studies of very advanced facilities is a wise strategy.

Sincerely yours,

Herbert Friedman, *Chairman*
Assembly of Mathematical
and Physical Sciences

Astronomy Survey Committee

GEORGE B. FIELD, Harvard-Smithsonian Center for Astrophysics, *Chairman*
MICHAEL J. S. BELTON, Kitt Peak National Observatory
E. MARGARET BURBIDGE, University of California, San Diego
GEORGE W. CLARK, Massachusetts Institute of Technology
S. M. FABER, University of California, Santa Cruz
CARL E. FICHTEL, NASA Goddard Space Flight Center
ROBERT D. GEHRZ, University of Wyoming
EDWARD J. GROTH, Princeton University
JAMES E. GUNN, Princeton University
DAVID HEESCHEN, National Radio Astronomy Observatory
RICHARD C. HENRY, The Johns Hopkins University
RICHARD A. McCRAY, Joint Institute for Laboratory Astrophysics and the University of Colorado
JEREMIAH OSTRIKER, Princeton University
EUGENE N. PARKER, University of Chicago
MAARTEN SCHMIDT, California Institute of Technology
HARLAN J. SMITH, University of Texas, Austin
STEPHEN E. STROM, Kitt Peak National Observatory (*ex officio*)
PATRICK THADDEUS, NASA Goddard Institute for Space Studies and Columbia University
CHARLES H. TOWNES, University of California, Berkeley
ARTHUR B. C. WALKER, Stanford University
E. JOSEPH WAMPLER, University of California, Santa Cruz

PAUL BLANCHARD, *Executive Secretary*
DALE Z. RINKEL, *Administrative Secretary*

Preface

By the late 1970's, rapid advances in astronomical research had established the necessity for a new survey of the needs of astronomy and astrophysics to follow that presented in the Greenstein report, published in 1972. Following the appointment of the Chairman of a proposed new Astronomy Survey Committee in April 1978, preliminary planning began for the two-year study that has produced the present report, *Astronomy and Astrophysics for the 1980's.*

In the summer of 1978, the Chairman wrote to 223 astronomers and physicists across the nation, asking for their nominations for membership on the new Committee; more than half responded, suggesting the names of 229 different scientists. From this list, seven scientists were selected to join the Chairman as the nucleus of the Committee, which first met in December 1978. Through the addition of other experts in various fields of astronomy and astrophysics, the Survey Committee eventually grew to 21 members, representing a wide variety of institutions in various parts of the country. Its final meeting was held in December 1980.

The Academy charged the Committee with the development of priorities for a comprehensive program in astronomy and astrophysics for the 1980's. This task proved highly challenging, as the scope and power of astronomical techniques have grown impressively since the completion of the Greenstein report. For example, x-ray astronomy is no longer an emerging field of research occupied with studies of only a few bizarre objects; it has now matured to stand with optical and radio astronomy as a comprehensive and versatile area of astrophysical inquiry. Infrared astronomy has grown from infancy to youthful vigor; ultraviolet, gamma-ray, and cosmic-ray observations have contributed important new results; and the more established fields of optical and radio astronomy have themselves recorded dramatic advances during the past ten years. To these achievements must be added the contributions to astronomy made by the separate program of lunar and planetary exploration using deep-space probes, together with new results from related sciences. The overall charge to the present Astronomy Survey Committee thus represented a challenge even greater than that so well met by the Greenstein Committee in its study of this rapidly changing field.

In order to carry out its charge, the Committee first had to delineate its domain of inquiry. For the purposes of this report, observational astronomy is taken to be the obtaining of information about astronomical bodies by remote sensing from the surface of the Earth, from the Earth's atmosphere, and from Earth orbit in space. The Astronomy Survey Committee has therefore *not* considered recommendations for:

1. Instruments or facilities on spacecraft designed to escape Earth orbit, such as deep-space probes to the planets, Sun, or other bodies in the solar system;
2. Programs to study the nature or environment of the Earth itself, such as the Earth's atmosphere or magnetosphere;
3. The gathering and laboratory analysis of samples of matter originating beyond the Earth; or
4. Instruments or facilities intended primarily to test the predictions of different theories of gravitation, rather than to obtain information about astronomical bodies, whether or not such tests are conducted in space or involve celestial observations.

The foregoing restrictions on the present study are not meant in any way to minimize the importance of the areas of science thereby excluded—namely, solar-system exploration by deep-space probes, Earth sciences, meteoritics, and gravitational physics. Each of these scientific areas, although not the subject of recommendations in this report, is a vigorous subject of research in its own right; each is also the province of one or more appropriately constituted committees or advisory groups, charged, like the Astronomy Survey Committee, with the formulation of recommendations for its guidance.

Our national program of planetary exploration has been ably guided by strategies developed by the Committee on Planetary and Lunar Exploration (COMPLEX) of the Space Science Board (SSB), which has published a series of reports dealing with strategies for the exploration of the outer solar system, the inner planets, and primitive solar-system bodies. Later in this decade, a probe to the Sun should become feasible; such a mission has been recommended by the SSB's Committee on Solar and Space Physics (CSSP), which has also considered the role of *in situ* measurements in the study of the solar wind and its interaction with planetary atmospheres, ionospheres, and magnetospheres. Eventually, deep-space probes devoted entirely to astronomical objectives should be considered; an identification of their programmatic roles may emerge through future recommendations of SSB's Committee on Space Astronomy and Astrophysics (CSAA). Investigations of the Earth's atmosphere and magnetosphere by space techniques have an important bearing on our knowledge of the structure and history of the inner solar system; strategies for such investigations have been developed by CSSP and by the SSB's Committee on Earth Sciences (CES). Through studies of meteorites and samples of interplanetary dust, it has been recognized that such material holds vital clues to the early history and chemical composition of the solar system, particularly through preservation of isotope ratios of astrophysically important elements; a thorough discussion of meteoritics and related research areas is presented in the recent report by COMPLEX on primitive solar-system bodies. Finally, the Committee notes that the study of gravitational physics is of particular importance to astronomy, as the large-scale structure of the Universe is determined almost entirely by the action of gravitational forces. Some of the most important future tests of the General Theory of Relativity and other theories of gravitation may require experiments in space; the recent report of the SSB's *ad hoc* Committee on Gravitational Physics presents a strategy for space research in gravitational physics for the 1980's.

To begin the present survey of astronomy and astrophysics—and particularly to help identify the most important scientific questions for the 1980's—the Committee first established seven Working Groups. Four of these were chosen according to objects or physical regions of scientific study: Solar Physics, Planetary Science, Galactic Astronomy, and Extragalactic Astronomy. A fifth Working Group, on Related Areas of Science, was chosen to survey developments during the 1970's in other sciences of importance to astronomy and to bring to the Committee's attention the directions such research might take in the coming decade. The final two Working Groups, on Astrometry and on the Search for Extraterrestrial Intelligence (SETI), were established to examine the rather different opportunities presented by these two specialized fields. The reports of the Working Groups served as internal Committee documents; because of their quality and vision the Committee has sponsored their publication as an independent supplement to the present report. These reports—which reflect the conclusions of the Working Groups rather than recommendations of the Survey Committee—are to be found in the companion volume, *Challenges to Astronomy and Astrophysics: Working Documents of the Astronomy Survey Committee.*

To aid the Committee in the formulation of recommendations, the Committee also established six Panels. Five of these correspond to techniques of astronomical study and were intended to provide direct channels of communication with the communities of scientists employing them: High Energy Astrophysics; Ultraviolet, Optical, and Infrared Astronomy; Radio Astronomy; Theoretical and Laboratory Astrophysics; and Data Processing and Computational Facilities, although the last has obvious connections to the rest. A sixth Panel, on Organization, Education, and Personnel, was charged with a more general investigation of the health of the professions involved in astronomical research. All the Panels were asked to make recommendations in their respective areas for the consideration of the Survey Committee; these may be found in the second volume of the present study, *Astronomy and Astrophysics for the 1980's, Volume 2: Reports of the Panels* (National Academy Press, Washington, D.C., 1982).

The Committee sought from the beginning to engage the widest possible participation of the scientific community. Together with consultants, the Working Groups, Panels, and the Committee itself involved more than 130 scientists. Meetings of nearly all of these Survey groups were held at centers of astronomical research across the country in order to provide opportunities for other scientific colleagues to be heard. In Open Letters of April 1979 and November 1979 to the 3700 members of the American Astronomical Society (AAS), the Committee Chairman explained the organization of the Survey, listed its participants, and invited the comments and suggestions of the scientific community. The second Open Letter was accompanied by a form that could be used by AAS members and their colleagues to propose astronomy projects or programs for the 1980's. The many responses received from this second mailing were directed to the appropriate Panels for evaluation and, where appropriate, incorporation into the Panel reports and recommendations.

The Committee also benefited from the study of numerous other reports dealing with future planning for astronomy and astrophysics. One of those most directly relevant to the present Survey is the recent report, *A Strategy for Space Astronomy and Astrophysics for the 1980's* (National Academy of Sciences, Washington, D.C., 1979), produced by the CSAA of the SSB. Although restricted to research from space, this study went far in delineating many of the issues later faced by the Committee, and it also suggested the format of the present Committee's recommendations. The Committee believes that its own recommendations and those of CSAA are consistent and mutually supportive in the area of space astronomy and astrophysics.

The recommendations for a program in astronomy and astrophysics, summarized in Chapter 2, are presented in three groups: Prerequisites for New Research Initiatives, New Programs, and Programs for Study and Development. The major criteria for selection of scientific programs were scientific importance and technological readiness, although consideration was also given to cost-effectiveness. The recommendations of the Panels figured heavily in the debate over selection and priorities, but the fact of a Panel recommendation was not considered in itself sufficient for high priority or even inclusion in the recommended program. The final recommendations are those of the Survey Committee alone.

Chapter 3 of the present report, "Frontiers of Astrophysics," presents a discussion of a number of major areas of astronomical research today and illustrates the excitement of this field, which, perhaps more than any other science, has so captured the attention and admiration of the general public; it also provides scientific background for the recommended program. The remaining chapters of the report describe the recommended program in greater detail and discuss the importance of each recommendation to the future development of astronomy and astrophysics in the United States.

The Astronomy Survey Committee is grateful for the contributions of many individuals to its effort, particularly the members of the Panels and Working Groups listed in Appendix C. In addition, the Committee wishes to thank the following: Bruce Gregory and Hope Bell of the National Research Council staff, for invaluable advice and assistance; William E. Howard III of the National Science Foundation, and Adrienne Timothy, Jeffrey Rosendhal, and Franklin Martin of the National Aeronautics and Space Administration, for helpful discussions on many occasions and other much appreciated support; Ivan King, former President of the American Astronomical Society, Peter Boyce, AAS Executive Officer, and other members of the AAS

Council, for their interest and help in publicizing Survey activities; and, finally, Paul Blanchard, Executive Secretary; Dale Rinkel, Administrative Secretary; and Martha H. Liller for their tireless efforts on behalf of the Astronomy Survey generally.

George B. Field, *Chairman*
Astronomy Survey Committee

Contents

Astronomy
and Astrophysics
for the 1980's

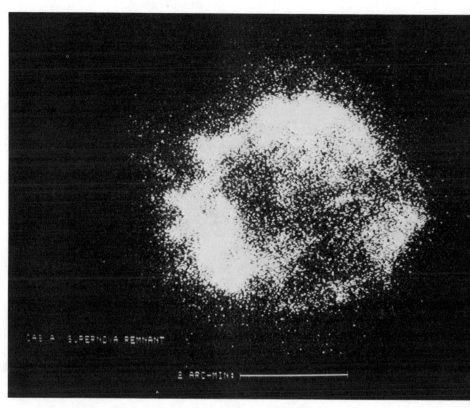

Einstein *Observatory x-ray image of the expanding supernova remnant Cassiopeia A. (Photo courtesy of S. S. Murray, Harvard-Smithsonian Center for Astrophysics)*

1

Introduction

Nature offers no greater splendor than the starry sky on a clear, dark night. Silent, timeless, jeweled with the constellations of ancient myth and legend, the night sky has inspired wonder throughout the ages. It is a wonder that leads our imagination far from the confines of Earth and the pace of present day, out into boundless space and cosmic time itself.

Astronomy, born in response to that wonder, is sustained by two of the most fundamental traits of human nature: the need to explore and the need to understand. Through the interplay of discovery, the aim of exploration, and analysis, the key to understanding, answers to questions about the Universe have been sought since the earliest times, for astronomy is the oldest of the sciences. Yet it has never been, since its beginnings, more vigorous or exciting than it is today.

Through modern astronomy, we now know that we are connected to distant space and time not only by our imagination but also through a common cosmic heritage: the chemical elements that make up our bodies were created billions of years ago in the hot interiors of remote and long-vanished stars. Their hydrogen and helium fuel finally spent, these giant stars met death in cataclysmic supernova explosions, scattering afar the atoms of heavy elements synthesized deep within their cores. Eventually

3

this material collected into clouds of gas in interstellar space; these, in turn, slowly collapsed to give birth to a new generation of stars. In this way, the Sun and its complement of planets were formed nearly 5 billion years ago. Drawing upon the material gathered from the debris of its stellar ancestors, the planet Earth provided the conditions that ultimately gave rise to life. Thus, like every object in the solar system, each living creature on Earth embodies atoms from distant corners of our Galaxy and from a past thousands of times more remote than the beginnings of human evolution.

Although ours is the only planetary system we know, others may surround many of the hundreds of billions of stars in our Galaxy. Elsewhere in the Universe, beings with an intelligence surpassing our own may also at this moment gaze in wonder at the night sky, impelled by an even more powerful imagination. If such beings exist—possibly even communicating across the vast expanses of interstellar space—they, too, must share our cosmic heritage.

This recognition of our cosmic heritage is a relatively recent achievement in astronomy. However, it is but one of many such insights that our generation alone has been privileged to attain. In all of history, there have been only two periods in which our view of the Universe has been revolutionized within a single human lifetime. The first occurred three and a half centuries ago at the time of Galileo; the second is now under way.

EXPLORATION AND UNDERSTANDING

The discoveries of the past 20 years, made from both ground-based and space observatories, have radically changed our concepts of the origin and evolution of stars, galaxies, and the Universe itself.

The 1960's saw the discovery of quasars, x-ray sources outside the solar system, the cosmic microwave background radiation, pulsars, high-energy celestial gamma rays, large-scale inhomogeneities in the solar corona, and polyatomic molecules in interstellar clouds. The rapid pace of discovery continued in the 1970's:

1970: *Uhuru*, the first satellite x-ray observatory, is launched; it reveals that many bright Galactic x-ray sources are neutron stars accreting matter from nearby companion stars and that many clusters of galaxies are pervaded by hot intergalactic gas whose mass rivals that of the galaxies themselves.

1971: Using very-long-baseline interferometry (VLBI), radio astronomers find that in a number of quasars, individual radio components appear to move apart from one another at speeds greater than that of light; there is still no definitive explanation for this phenomenon.

1972: The *Copernicus* satellite is launched to provide high-resolution ultraviolet spectroscopic observations of stars and interstellar gas; vast regions of the interstellar medium are shown to be heated to hundreds of thousands of degrees by the shock waves from supernova explosions.

1972: The Gamma Ray Explorer satellite is launched, providing the first detailed picture of the Galactic plane in gamma rays, as well as measurements of the spectrum of the diffuse extragalactic gamma radiation; two pulsars are also detected.

1973: Ground-based observations reveal a cloud of sodium gas surrounding Jupiter's satellite Io; later a torus of gas injected by Io into orbit around Jupiter is found, presaging the discovery of Io's vigorous volcanic activity by the *Voyager* 1 spacecraft.

1973: Observations by the Skylab satellite confirm earlier indications that high-velocity solar-wind streams flow from "coronal holes," solar regions of much reduced x-ray and ultraviolet emission; these streams are later recognized as the source of many geomagnetic disturbances on the Earth.

1974: Radio observations of the pulsar PSR 1913 + 16 show that it orbits a companion star every 6 hours; precise measurements of this binary pulsar later confirm the prediction of the General Theory of Relativity that a binary-star system radiates energy in the form of gravitational waves.

1975: Observations confirm theoretical predictions that the solar surface undergoes 5-minute oscillations as a result of waves in the solar interior; detailed measurements of these oscillations later provide seismic probes of solar structure that reach nearly to the Sun's core.

1976: An experiment to measure the flux of neutrinos from the center of the Sun, in progress since 1970, shows that the flux is no more than one third of that predicted by current theory; this finding prompts a searching re-examination of the theory of generation of nuclear energy in stars.

1977: In the first such discovery since the rings of Saturn were discovered in the seventeenth century, rings are found around

Uranus. In 1979, *Voyager* 1 detects the first known ring around Jupiter and in 1980 reveals complex structure in the rings of Saturn.

1978: The *Einstein* Observatory, carrying an imaging x-ray telescope, is launched to acquire the first high-resolution images and detailed spectra of the x-ray sky; these observations provide important new insights into the properties of nearly every kind of astronomical object.

1978: Infrared spectroscopy reveals the expansion of a shell of dust and gas away from a protostellar condensation in the Orion nebula; if the expansion is due to the pressure of radiation from a newborn star within the shell, this star must be only 2000 years old, and thus the youngest ever observed.

1978: Radio and x-ray emissions from the Galactic object SS 433 prompt optical studies that reveal that it emits jets of matter at about one fourth of the speed of light; SS 433 may be a small-scale stellar counterpart of the mysterious objects that produce energetic jets in some quasars.

1978: The International Ultraviolet Explorer (IUE) satellite is launched; observations establish temperature scales for hot stars, reveal resemblances between quasars and Seyfert galaxies, and show that mass loss from stars at rates high enough to affect stellar evolution is nearly universal.

1979: Isotopic analysis of cosmic rays shows conclusively that they constitute a sample of matter with a nucleosynthetic history different from that of the Sun.

1979: The most intense gamma-ray burst ever recorded is observed by an international network of spacecraft; it is shown to have come from the same direction as that of a supernova remnant in the Large Magellanic Cloud.

1979: A multiple quasar is detected and shown to be the result of the splitting of light beams from a single, distant quasar by the "gravitational lens" effect of an intervening galaxy; a second example is found in 1980.

These and other key discoveries of the past 20 years must certainly be ranked in significance with those in the decades following Galileo's telescopic observations. They are the direct results of the imaginative application of new technologies for obtaining and recording astronomical information. Indeed, the pace of discovery has been set largely by the rate of advance of technology.

During the 1970's, for example, the maturing of space technology and instrumentation opened up the far-ultraviolet, x-ray, and gamma-ray regions of the spectrum to extensive new observations. Infrared astronomy from the ground and from aircraft and balloons made major strides, with orbiting telescopes scheduled for the decade ahead. The development of highly efficient detectors for optical astronomy permitted the study of extremely faint and previously inaccessible objects with large telescopes and enabled smaller facilities to become major research tools. New technologies for telescope construction—employing thin mirrors or mirror segments, ultrashort focal lengths, lightweight support structures, and innovative designs for domes or housings—made it both technically and economically feasible to build much larger telescopes for optical and infrared studies. Computers evolved into workaday tools in both the laboratory and the observatory so that it is now feasible to apply them to data reduction, image processing, and theoretical studies in research facilities across the nation.

The 1980's will almost certainly witness further discoveries to rank with those of recent decades; however, new discoveries do not necessarily lead immediately to a deeper understanding of the Universe. Such understanding, the ultimate goal of astronomy, requires the careful analysis of observations to test their validity and to assess their relationship to currently accepted knowledge.

In contrast to observational discoveries, major advances in our understanding of the Universe seldom burst suddenly upon the scientific world, bearing key dates that will ring through history. Rather, they usually require the systematic measurement, classification, interpretation, and study of many objects and phenomena, whose common properties and unifying features may not become apparent until years of effort have been devoted to the task. Our empirical knowledge of the Universe is gained almost entirely from the laborious, time-consuming collection of the feeble electromagnetic radiations that reach us from very distant objects. Instruments and facilities to follow up and analyze initial discoveries must therefore be more powerful, more versatile, and longer lived than those intended primarily for discovery. They must also be able to return and process new information at more rapid rates, if advances in our understanding are not to lag behind the pace of initial exploration.

The gathering of observational data is, however, only the first step in the quest for understanding; these data must then be analyzed and interpreted with the aid of astrophysical theory. The task of theory is to develop physical models for the mechanisms that underlie the observed phenomena, to calculate the properties of the models, to test these properties through comparisons with the observations, and to predict the results of new observations. When such comparisons are favorable and the predictions accurate, our understanding has advanced. The construction of astrophysical theories is based on the results from mathematics and the other sciences, particularly physics and chemistry. Laboratory studies are essential to provide the atomic, molecular, and nuclear data needed in theoretical analysis. All of these activities essential to the search for understanding are nurtured in our universities and research institutions, where the astronomical knowledge and skills gained in the past are refined by new research results and transmitted to those who will follow.

The past decade has seen impressive progress toward understanding many of the more significant and challenging problems of contemporary astronomy. For example, in the decade between the launch of the *Uhuru* satellite in 1970 and the imaging and spectroscopic work of the *Einstein* Observatory, x-ray astronomy progressed from a low-resolution survey of the x-ray sky that revealed a few hundred mostly unidentified sources, to the detailed study of x-ray emission from thousands of astronomical objects of all classes, ranging from normal stars to quasars and clusters of galaxies. The model of x-ray stars based on the theory of accretion of matter onto compact objects such as neutron stars and black holes has provided a possible model for also understanding the vastly more powerful energy sources of active galactic nuclei and quasars.

Another example of impressive progress recorded during the 1970's is our increased understanding of the structure and energy balance of the interstellar medium. Ultraviolet spectroscopy provided by the *Copernicus* and IUE satellite observatories, coupled with x-ray observations from rockets and satellites, has revealed that large regions of interstellar space are filled with gas heated to hundreds of thousands of degrees by shock waves from supernova explosions; other ultraviolet studies have shown that mass loss is a feature of normal stellar evolution. These investigations, powerfully aided by theoretical and laboratory studies of atomic and molecular properties, have led to a new understanding of the ways in which stars are coupled to their surrounding environment.

In some important areas of astronomy, however, understanding has progressed less rapidly, often because critical observations needed to distinguish between alternative theoretical possibilities are lacking. One such area is star formation, now known to be occurring in certain regions of the disks of many galaxies, including our own. The phenomena involved are extremely complex. Stars are born in clouds of gas and dust that form at low temperatures and radiate predominantly in the infrared and millimeter wavelength regions of the spectrum. Despite the great improvements in detectors and ground-based instrumentation in these wavelength regions during the 1970's, a new generation of observational facilities both in space and on the ground will be required to furnish the increases in spatial and spectral resolution needed for detailed comparison of regions of star formation with increasingly sophisticated theoretical models.

The evolution of galaxies is another area in which our understanding has been impeded by the lack of observations of sufficient numbers of objects at or beyond the limits of present facilities, although the 1970's provided fascinating hints of what we will eventually find. The 1960's furnished clear evidence that we live in an evolving Universe: the detection of the cosmic microwave background radiation, believed to be a relic of the big bang, and the discovery that the densities of radio sources and quasars increase with distance as one looks back to earlier stages in the history of the Universe. Recent observations suggest that distant galaxies are bluer than nearby ones, providing the first direct evidence for the evolution of galaxies. A better understanding of the evolution of the Universe will require measurements of much more distant and hence much fainter galaxies than are currently observable with our largest telescopes equipped with the best detectors. Only through such measurements will we be able to determine the general properties of galaxies at early stages in their evolution, when these properties were noticeably different from those of galaxies closer to us.

A DECADE OF OPPORTUNITY

The Astronomy Survey Committee believes that the programs recommended in this report are those that will most effectively advance both exploration and understanding during the coming decade.

These programs address the most significant questions that confront contemporary astronomy. For example: What is the large-scale structure of the Universe? How do galaxies evolve? What role do violent events play in the evolution of the Universe? How are stars and planets formed? What causes activity on the surfaces of the Sun and other stars? How widespread are life and intelligence in the Universe? And do the connections between astronomy and the nature of the fundamental forces hold the key to a unified understanding of all cosmic processes? Chapter 3 presents a detailed examination of these questions and a discussion of how they may be addressed through implementation of the Committee recommendations.

The technological developments of the 1970's permit these problems to be attacked now, with a high degree of effectiveness, and at reasonable cost. The maturing of space astronomy, advances in detectors, reduction in cost of computers, and new technologies for constructing telescopes have already been mentioned; to these developments must be added more specialized advances, such as the refinement of radio VLBI; short-wavelength antenna design and fabrication; new approaches to the detection of cosmic rays, neutrinos, and gravitational waves; and infrared interferometry. New facilities incorporating these developments will undoubtedly yield advances in our understanding of the Universe and, like those of the recent past, will produce additional new discoveries of their own.

One should also recognize the importance of astronomical research as an essential contribution to the scientific and technological vigor of the nation. The entire field has grown explosively during the past two decades, and the research activity of trained astronomers has become a significant component of the U.S. scientific enterprise. Maintaining U.S. leadership in astronomical research will be an increasing challenge during the 1980's, as other nations continue to develop their scientific capabilities.

Research in astronomy has furthermore often had a surprising impact on technology. It was through attempts to understand the orbiting of the Moon and planets that Newton discovered the laws of motion that are the basis of modern engineering. The theory of radiative transfer, the development of which was stimulated by the need of astrophysicists to understand the escape of radiation from the outer layers of a star, was later applied by engineers to studies of the escape of neutrons from nuclear reactors. In attempting to explain the source of stellar energy, astrophysicists helped to advance our understanding of thermonuclear fusion; astrophysical studies of plasmas in space have contributed to the theory of magnetic plasma confinement, which is applied to the control of thermonuclear fusion for practical purposes.

New concepts emerge as astronomy attempts to understand the Universe. Picked up and developed by other scientists, they become part of the knowledge that underlies modern technology. Today astronomers are again confronted by puzzling phenomena in space, such as relativistic beams of matter, interstellar masers, and superdense matter. The new concepts that will be required to understand such phenomena will surely have an impact on the technology of the future.

The programs recommended are those that the Committee believes are most likely to promote scientific knowledge in the coming decade. However, no projections for a decade of research can be exhaustive, particularly for an age of discovery like the one in which we live. A different and still more wonderful view of the cosmos will almost surely be revealed to us in the years ahead. The discoveries of our generation have brought us to the threshold of a revolution in physical thought, in which the properties of elementary particles may hold the key to understanding the early history of the Universe and in which the quantum properties of gravitation, unrecognized in the theories of Newton and Einstein, may play a central role in understanding cosmic evolution. Flexibility and openness to new opportunities, as well as implementation of the programs recommended here, will be needed to respond effectively to discoveries and developments that cannot now be foreseen.

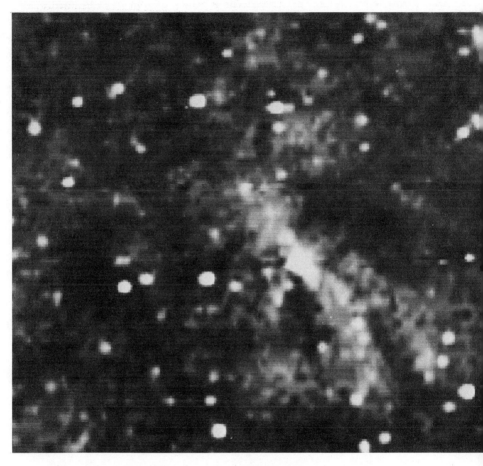

A 2-μm infrared image of the center of the Galaxy, hidden
by 30 magnitudes of optical extinction. (Photo courtesy of G.
Neugebauer and E. Becklin, California Institute of
Technology)

2

Recommended Priorities for Astronomy and Astrophysics in the 1980's

The Astronomy Survey Committee takes note at the outset of the support provided to U.S. astronomy and astrophysics over the past decades through the scientific programs of the National Science Foundation (NSF), the National Aeronautics and Space Administration (NASA), and other federal agencies. This support has enabled U.S. astronomical research to maintain an overall position of world leadership and has vastly widened our horizons for exploration of the Universe.

The programs recommended in this report have been selected from research activities that were, at the beginning of the Survey, candidates for implementation in fiscal year 1983 and beyond. Before presenting a summary of its recommendations, however, the Committee wishes to emphasize the importance of approved, continuing, and previously recommended programs to the progress of astronomical research during the remainder of the decade. The present Committee's recommendations take explicit account of such programs and build upon them.

The Committee calls particular attention to the need for support of the following approved and continuing programs, for which the order of listing carries no implication of priority: Space Telescope and the associated Space Telescope Science Institute; second-generation Space Telescope instrumentation; the Gamma Ray Observatory; NASA level-of-effort observational programs, including research with balloons, aircraft, and sounding rockets, together with the Ex-

13

plorer and Spacelab programs; the Solar Optical Telescope and the Shuttle Infrared Telescope Facility for Spacelab; facilities for the detection of neutrinos from the solar interior; federal grants in support of basic astronomical research at U.S. universities; and programs at the National Astronomy Centers. The 25-Meter Millimeter-Wave Radio Telescope, which was recommended in an earlier form in the Greenstein report, has not yet been implemented. The present status of these approved, continuing, and previously recommended programs is described later in this chapter; their importance for the health of U.S. astronomy in the 1980's is discussed in Chapter 4.

SUMMARY OF THE RECOMMENDED PROGRAM

The Astronomy Survey Committee recommendations for a program in astronomy and astrophysics for the 1980's fall into three general categories:

Prerequisites for new research initiatives;
New programs; and
Programs for study and development.

As noted in the Preface, the observational components of these recommendations are restricted to remote sensing from the Earth or its vicinity. A background and overview of the recommendations follows later in this chapter.

Prerequisites for New Research Initiatives

In order to be effective, the recommended new research initiatives for the coming decade must be supported by a set of Prerequisites that apply to both the gathering and the analysis of the data produced. These Prerequisites are essential for the success of major programs but are inexpensive by comparison. Although significant support already exists for each, the Committee strongly recommends substantial augmentations in the following areas, in which the order of listing carries no implication of priority:

A. *Instrumentation and detectors*, to utilize the latest technology to enhance the efficiency of both new and existing telescopes in the most cost-effective manner;

B. *Theory and data analysis*, to facilitate the rapid analysis and understanding of observational data;

C. *Computational facilities*, to promote data reduction, image processing, and theoretical calculations;

D. *Laboratory astrophysics*, to furnish the atomic, molecular, and nuclear data essential to the interpretation of nearly all astronomical observations; and

E. *Technical support at ground-based observatories*, to ensure that modern astronomical instrumentation is maintained in the best condition permitted by the state of the art.

A detailed consideration and justification of these Research Prerequisites appears in Chapter 5.

New Programs

The Astronomy Survey Committee recommends the approval and funding of new programs in astronomy and astrophysics for the 1980's. These have been arranged into three categories according to the scale of resources required.

A. *Major New Programs* The Committee believes that four major programs are critically important for the rapid and effective progress of astronomical research in the 1980's and is unanimous in recommending the following order of priority:

1. *An Advanced X-Ray Astrophysics Facility* (AXAF) operated as a permanent national observatory in space, to provide x-ray pictures of the Universe comparable in depth and detail with those of the most advanced optical and radio telescopes. Continuing the remarkable development of x-ray technology applied to astronomy during the 1970's, this facility will combine greatly improved angular and spectral resolution with a sensitivity up to one hundred times greater than that of any previous x-ray mission.

2. *A Very-Long-Baseline* (VLB) *Array* of radio telescopes designed to produce radio images with an angular resolution of 0.3 milliarcsecond. Among many potential applications of profound importance, this instrument will probe the small-scale structure surrounding the enigmatic energy sources in the cores of quasars and active galactic nuclei and will directly determine the distance scale within our Galaxy with unprecedented accuracy.

3. *A New Technology Telescope* (NTT) of the 15-m class operating from the ground at wavelengths of 0.3 to 20 μm, to provide a tenfold increase in light-gathering capacity at visual wave-

lengths and a hundredfold increase in speed for spectroscopy at infrared wavelengths, with application to a very wide range of scientific problems. The Committee finds the scientific merit of this instrument to be as high as that of any other facility considered and emphasizes that its priority ranking does not reflect its scientific importance but rather its state of technological readiness. *The design studies needed before NTT can be constructed are of the highest priority and should be undertaken immediately.*

4. *A Large Deployable Reflector* in space, to carry out spectroscopic and imaging observations in the far-infrared and submillimeter wavelength regions of the spectrum that are inaccessible to study from the ground, thus extending the powerful capabilities of NTT to these longer wavelengths. Such an instrument, in the 10-m class, will present unprecedented opportunities for studying molecular and atomic processes that accompany the formation of stars and planetary systems.

B. *Moderate New Programs*   In rough order of priority, these are:

1. *An augmentation to the* NASA *Explorer program*, which remains a flexible and highly cost-effective means to pursue important new space-science opportunities covering a wide range of objects and nearly every region of the electromagnetic spectrum.

2. *A far-ultraviolet spectrograph in space*, to carry out a thorough study of the 900–1200-Å region of the spectrum, important for studies of stellar evolution, the interstellar medium, and planetary atmospheres.

3. *A space* VLB *interferometry antenna* in low-Earth orbit, to extend the powerful VLBI technique into space in parallel with the rapid completion of a ground-based VLB Array, in order to provide more detailed radio maps of complex sources, greater sky coverage, and higher time resolution than the Array can provide alone.

4. *The construction of optical/infrared telescopes in the 2–5-m class*, to observe transient phenomena, conduct long-term survey and surveillance programs, provide crucially needed ground-based support to space astronomy, and permit the development of instrumentation under realistic observing conditions. The Committee particularly encourages federal assistance for those projects that will also receive significant nonfederal funding for construction and operation.

5. *An Advanced Solar Observatory* in space, to provide observations of our Sun—the nearest star—simultaneously at optical, extreme ultraviolet, gamma-ray, and x-ray wavelengths, to carry out long-term studies of large-scale circulation, internal dynamics, high-energy transient phenomena, and coronal evolution.

6. *A series of cosmic-ray experiments* in space, to promote the study of solar and stellar activity, the interstellar medium, the origin of the elements, and violent solar and cosmic processes.

7. *An astronomical Search for Extraterrestrial Intelligence* (SETI), supported at a modest level, undertaken as a long-term effort rather than as a short-term project, and open to the participation of the general scientific community.

C. *Small New Programs*   The program of highest priority is:
- An antenna approximately 10 m in diameter for submillimeter-wave observations, at an excellent ground-based site.

Other programs of outstanding scientific merit, in which the order of listing carries no implication of priority, are as follows:
- A spatial interferometer for observations of high angular resolution in the mid-infrared region of the spectrum;
- A program of high-precision optical astrometry; and
- A temporary program to maintain scientific expertise at U.S. universities during the 1980's through a series of competitive awards to young astronomers.

Detailed discussion and justification of the New Programs appears in Chapter 6.

Programs for Study and Development

Planning and development are often time-consuming, especially for large projects. It is therefore important during the coming decade to begin study and development of programs that appear to have exceptional promise for the 1990's and beyond. Projects and study areas recommended by the Committee in this category include the following, in which the order of listing carries no implication of priority:

A. Future x-ray observatories in space;

B. Instruments for the detection of gravitational waves from astronomical objects;

C. Long-duration spaceflights of infrared telescopes cooled to cryogenic temperatures;

D. A very large telescope in space for optical, ultraviolet, and near-infrared observations;

E. A program of advanced interferometry in the radio, infrared, and optical spectral regions;

F. Advanced gamma-ray experiments; and

G. Astronomical observatories on the Moon.

Detailed discussion of the Programs for Study and Development appears in Chapter 7.

ESTIMATED COST OF THE RECOMMENDATIONS

In order to establish the overall scale of the recommended total program, the Committee gives in Table 2.1 its own approximate estimates of the requirements for new funding over the next 10 years in millions of 1980 dollars. Funds for projects to be supported by NASA represent research-and-development funds within NASA's Office of Space Science and Applications (OSSA); funds for projects to be supported by NSF represent total cost to NSF. Operating costs are included for those facilities expected to become operational in the 1980's.

The funding entries for the Prerequisites for New Research Initiatives represent augmentations to the present levels of support for these activities within NSF and NASA. As it is expected that the two agencies will work together to coordinate support for the Prerequisites, specific agency responsibility is not indicated in the following table. However, since the Prerequisites provide support to space- and ground-based research at comparable levels, the Committee anticipates that the funding augmentations to be provided by NASA and NSF will be roughly equal in magnitude.

In the cases of the New Programs, the division between space- and ground-based projects is clear. Funds listed for the Explorer program represent an augmentation to NASA's level-of-effort budget for that program; the operations costs listed for ground-based projects, together with the temporary program to maintain scientific expertise at U.S. universities, represent further augmentations to the operations budget of NSF's Astronomy Division. Remaining New Program costs represent new-funding requirements for either NASA (new starts within OSSA) or NSF (major construction within the Astronomy Division).

TABLE 2.1 Requirements for New Funding

	(Millions of 1980 Dollars)
PREREQUISITES FOR NEW RESEARCH INITIATIVES	
A. Instrumentation and detectors (doubling of present $15 million/year level of effort by increments over 10 years)	$ 75
B. Theory and data analysis (augmentation by $5 million/year)	50
C. Computational facilities (30 minicomputer systems installed at a rate of 5 systems/year, including operations)	20
D. Laboratory astrophysics (augmentation by $2.5 million/year)	25
E. Technical support at ground-based observatories, including 40 new support positions	20
DECADE TOTAL, PREREQUISITES	$190
NEW PROGRAMS	
A. *Major New Programs* In order of priority:	
1. Advanced X-Ray Astrophysics Facility (AXAF)	$500
2. Very-Long-Baseline (VLB) Array (including $15 million for operations)	50
3. New Technology Telescope (NTT)	100
4. Large Deployable Reflector in space	300
Decade Subtotal	$950
B. *Moderate New Programs* In rough order of priority:	
1. Augmentation to Explorer satellite program	$200
2. Far-ultraviolet spectrograph in space	150
3. Space VLB interferometry antenna	60
4. Optical/infrared telescopes in the 2–5-m class	20
5. Advanced Solar Observatory in space	200
6. Cosmic-ray experiments	100
7. An astronomical Search for Extraterrestrial Intelligence (SETI)	20
Decade Subtotal	$750
C. *Small New Programs* Of highest priority:	
• 10-m submillimeter-wave radio antenna (including $2 million for operations)	$ 4
Other important programs:	
• Spatial interferometer for the mid-infrared (including $1 million for operations)	3
• High-precision optical astrometry program	3
• Temporary program to maintain scientific expertise at U.S. universities	10
Decade Subtotal	$ 20
DECADE TOTAL, NEW PROGRAMS	$1,720

The cost estimates for AXAF and the VLB Array were derived with the help of individual scientists participating in current studies and are based on reasonably complete rough designs. The actual cost of NTT, however, cannot be estimated until further studies indicate which of several alternative conceptual designs will be most cost-effective; the figure given in Table 2.1 is meant as a limit that the Committee recommends should not be substantially exceeded. The estimated cost of the Large Deployable Reflector in space is highly uncertain because instrumentation of this type has not yet been developed and launched. Most of the costs estimated for the Moderate New Programs should be reasonable approximations, as they are based on experience with previous instruments of a similar nature. The costs given for the augmentation to the NASA Explorer program and for SETI, however, should be regarded as target figures for the level of effort the Committee finds appropriate.

The total cost in new funding estimated for the Prerequisites and New Programs together is about $1.9 billion in 1980 dollars. By comparison, the Greenstein report (1972) recommended new programs with an estimated cost of $844 million in 1970 dollars, or approximately $1.7 billion in 1980 dollars, and most of those programs were in fact implemented. *The program recommended here for the 1980's is thus roughly comparable in scale with that actually carried out during the 1970's on the basis of the recommendations of the Greenstein report.*

The Committee wishes to emphasize, however, that the present recommendations will require substantial increases in the budget of the Astronomy Division of NSF, the agency primarily responsible for the support of ground-based astronomy. If, as anticipated, NSF will provide roughly half of the additional funds required for the Prerequisites for New Research Initiatives, an increase of about 30 percent in the Astronomy Division's operations budget over the real level of expenditures during the 1970's will be required for NSF to carry out its share of the recommended program over the next decade. Funds needed by NSF for major construction over the next 10 years will also be substantially higher than those expended during the 1970's, which saw the completion of only one major project, the Very Large Array, at a cost of $78 million. *The Astronomy Survey Committee believes that these increases in the NSF budget for ground-based astronomy are essential to maintain an effective partnership with space astronomy during the 1980's.*

BACKGROUND AND OVERVIEW

The Greenstein Report

The publication of *Astronomy and Astrophysics for the 1970's* (the Greenstein report) by the National Academy of Sciences in 1972 had a powerful impact on the development of U.S. astronomy and astrophysics during 1972–1982. The federal government on the whole responded positively to its recommendations, with the result that the facilities available to U.S. astronomers have enabled them to remain at the frontiers of research. Here we review the responses to the recommendations of that report and their impact on the progress of science.

Radio Astronomy and the VLA The highest-priority recommendation of the Greenstein report was the construction of a Very Large Array (VLA) radio telescope, together with increased support for smaller facilities. Funded by NSF, the VLA was constructed in stages during the 1970's and was formally dedicated in 1980; by far the largest and most complex ground-based astronomical facility established to date, the VLA was completed on schedule and within budget. VLA studies of radio sources are already having a large impact on both Galactic and extragalactic astronomy (see, for example, the cover of this report).

The recommended increase in funding for smaller radio-astronomical facilities did not materialize, however, nor has funding yet been provided for a recommended millimeter-wave radio telescope, then projected to have a diameter of 65 m and to be operable at wavelengths down to 3 mm. Since the publication of the Greenstein report, the study of interstellar molecules at millimeter wavelengths has yielded insight into the process of star formation; as the science has progressively moved to shorter wavelengths, there has now evolved a need for a smaller, more precisely figured telescope of 25-m diameter, still offering high sensitivity and spatial resolution but operable at wavelengths down to 1 mm. The recommendation for a large centimeter-wave antenna was not implemented, although existing facilities for observations at wavelengths longer than 1 cm have been maintained and in some cases upgraded.

Optical Ground-Based Astronomy The second-priority recommendation was for a variety of steps to enhance the capability available to U.S.

optical astronomers. A key proposal was the development and construction of a multiple-mirror telescope (MMT) with aperture equivalent to that of a conventional telescope in the 3.8–5.0-m range; this project was to be followed by the construction of a larger MMT, of 10–15-m aperture, if that proved feasible, or by a conventional telescope of 5-m aperture if it did not. A 4.5-m MMT has in fact been developed jointly by the Smithsonian Institution and the University of Arizona, becoming operational in 1979. However, neither a larger MMT nor a conventional 5-m telescope has been funded, with the result that the largest instrument available to U.S. optical astronomers is still the 5-m Hale telescope on Mt. Palomar, which went into operation 35 years ago. As a result, optical spectroscopy of the faintest galaxies and quasars discovered by radio and x-ray astronomers has not kept pace with new discoveries, even though these extremely distant objects are of great interest because of their bearing on the nature of cosmic evolutionary processes early in the history of the Universe.

The recommendation also called for equipping existing telescopes with advanced electronic detectors and controls. Although major progress was made in the development of such devices during the 1970's, they have so far been provided to only a few major observatories. The capability of instruments now available for use on most of the nation's optical telescopes still lags far behind state-of-the-art technology.

The Greenstein report furthermore recommended that three telescopes in the 2.5-m class be constructed for a variety of purposes. As none of these has been funded, all the nation's major optical telescopes are heavily oversubscribed.

Infrared Astronomy The third-priority recommendation called for an across-the-board increase in support for infrared astronomy, which at that time was beginning to demonstrate its great importance. Support has increased substantially, through the funding of two major ground-based infrared telescopes (the 2.3-m University of Wyoming Infrared Observatory reflector and the 3-m Infrared Telescope Facility operated by NASA on Mauna Kea), the Kuiper Airborne Observatory program, and a balloon program. An international Infrared Astronomy Satellite (IRAS), scheduled for launch in 1982, will carry out a comprehensive, far-infrared survey of the sky, as called for in the Greenstein report.

X-Ray and Gamma-Ray Astronomy A series of four High Energy Astronomical Observatories (HEAO's) was the fourth-priority recommendation. In 1973, however, the entire HEAO program was restructured to reduce costs, and only three HEAO's were actually flown. Although it had considerably smaller capability than originally planned, the *Einstein* (HEAO-2) Observatory obtained detailed images of cosmic x-ray sources for the first time, with dramatic implications for a broad range of astronomical research.

In gamma-ray astronomy, an important survey was carried out by the Small Astronomical Satellite-B, and further results were obtained by HEAO-3, as well as by balloons and rockets. The most sensitive gamma-ray survey of the sky to date has been carried out by the European Space Agency's COS-B satellite.

Other Programs The Greenstein report recommended five other programs, in addition to the four primary recommendations and the two additional radio telescopes mentioned above. A recommended increase in support for aircraft, balloons, and rockets did not occur. The Orbiting Solar Observatory program was not continued as recommended; however, major programs of solar observations were carried out on the Apollo Telescope Mount as part of the manned Skylab program and on the Solar Maximum Mission spacecraft, which carried instruments designed principally for the study of solar flares. Both programs provided important insights into the physical processes that occur in the solar atmosphere. Theoretical and laboratory astrophysics did not receive the recommended substantial increase, although summer institutes in theoretical astrophyics of the type suggested in the Greenstein report have been implemented at several locations at very low cost and with excellent results. The recommended new facilities for astrometry were not funded.

Recommendation 9 of the Greenstein report called for an expanded program of optical space astronomy, leading to the launch of a 3.0-m, diffraction-limited Large Space Telescope in the early 1980's. NASA chose instead to develop a somewhat smaller 2.4-m Space Telescope (ST) for launch in 1985 and, in the interim, to extend the lifetime of the OAO-3 (*Copernicus*) satellite through the 1970's and to launch the International Ultraviolet Explorer (IUE) in 1978. The high-resolution imaging proposed in Recommendation 9 remains to be realized with ST, although important ultraviolet data are currently being obtained with IUE.

Perspective on the Present Survey

The internal advisory committees of NSF and NASA—including NSF's Astronomy Advisory Committee and NASA's Space Science Advisory

Committee—provide effective channels of communication with the scientific community, and they therefore play a significant role in the year-to-year funding decisions faced by these agencies. In addition, since the publication of the Greenstein report, the extent of guidance to NASA on long-range planning for space astronomy has been expanded through the activities of the National Research Council's (NRC) Space Science Board (SSB), particularly through the scientific strategies recommended by SSB's Committee on Space Astronomy and Astrophysics (CSAA) and Committee on Solar and Space Physics (CSSP). The recommendations of these bodies have increasingly helped to shape the course of U.S. space astronomy. Their most recent reports, to which reference will be made below, are the CSAA document, *A Strategy for Space Astronomy and Astrophysics for the 1980's* (National Academy of Science, Washington, D.C., 1979), and the CSSP document, *Solar-System Space Physics in the 1980's: A Research Strategy* (National Academy of Sciences, Washington, D.C., 1980).

It is therefore not appropriate for a "decade review" such as the Astronomy Survey to prescribe rigidly all agency funding decisions for the next decade. To do so would be to ignore, on the one hand, the guidance available to NSF and NASA from their own internal advisory bodies, particularly in reaction to year-by-year changes in scientific opportunities or in the funding climate; and, on the other, the guidance for space-science strategy already available (and in some cases already in the course of implementation) from the National Academy of Sciences.

The present Committee concluded that it could best help to advance astronomical research by focusing its attention on a coordinated program of ground- and space-based astronomy that would respond to long-run scientific opportunities. With this in mind, the Committee has limited its recommendations to programs originally proposed for funding in fiscal year 1983 or later, taking account of the earliest time that the publication of the present report could influence the budgets of federal agencies. Projects that were, at the beginning of the survey in 1979, candidates for implementation in fiscal year 1982 or earlier were therefore not considered for inclusion in the Committee's recommendations. A number of important approved and continuing programs are also not explicitly recommended by the Committee, although they form the programmatic base from which the present recommendations proceed. The present status of such projects and programs is as follows:

• Space Telescope (ST) and the Space Telescope Science Institute (STScI). Recommended in the Greenstein report, ST is now under development and is scheduled for launch in 1985. As recommended in the Space Science Board study, *Institutional Arrangements for the Space Telescope* (National Academy of Sciences, Washington, D.C., 1976), NASA has established STScI to direct the ST scientific programs and to support ST users.

• Second-generation instrumentation for ST. In its 1979 report, CSAA recommended that ST focal-plane instruments be periodically upgraded in order to fulfill the great scientific potential of ST.

• The Gamma Ray Observatory. The 1979 CSAA report listed this facility as one of the five major space-astronomy missions of highest priority for the 1980's. The Gamma Ray Observatory is now an approved NASA program and is projected for launch in 1987–1988.

• NASA level-of-effort observational programs, including balloons, aircraft, and sounding rockets, together with the Explorer and Spacelab programs. Strong support for each of these programs was recommended by CSAA in its 1979 report, and other NRC committees have also recommended support for one or more component programs within this group.

• The Shuttle Infrared Telescope Facility (SIRTF) and the Solar Optical Telescope (SOT) for Spacelab. The 1979 CSAA report recommended these projects as the first two major astrophysics facilities for Spacelab; the 1980 CSSP report also supported SOT as a facility of high importance for solar physics. SOT, now under development within the Spacelab program, is projected for launch in 1987–1988, while SIRTF is under study by NASA.

• Facilities for the detection of neutrinos from the solar interior. An experiment described in the Greenstein report, employing chlorine detectors at underground sites, can now be supplemented by investigations employing alternative detector materials, particularly gallium.

• Federal grants in support of basic astronomical research at U.S. universities. Such funds, provided primarily through NASA's Research and Analysis program and the grants program of NSF's Astronomy Division, are vital to the health of U.S. astronomy.

• Programs at the National Astronomy Centers. Adequate support of these programs will permit the Centers to continue to provide an extensive range of observational facilities to a wide and diverse user community.

• The 25-Meter Millimeter-Wave Radio Telescope. Recommended in an earlier form in the Greenstein report but not yet funded, this facility has been proposed for the past 4 years as the new program of highest priority in the NSF Astronomy Division's long-range plan. Chapter 4 presents a detailed discussion of the importance of these approved, continuing, and previously recommended programs to the health of U.S. astronomy during the 1980's.

Assignment of Priorities to Research Needs

There is a subtle interplay between science and technology in the pursuit of astronomical research. In more mature areas, such as optical astronomy, the addressing of specific scientific questions requires enhanced instrumental capability. In less mature fields, such as gamma-ray astronomy, instruments made possible for the first time by new technology will open up new fields of knowledge, even though the questions we can now ask are less specific. The Committee considered both the scientific questions and the emergence of new technological capabilities as part of the process of assigning priorities.

Chapter 3, "Frontiers of Astrophysics," is based on the findings of the Working Groups, which were asked to review the scientific opportunities in a number of subfields of astronomy. With this material as background, the Committee considered recommendations that had come to it from the Panels, which had been asked to review possibilities for new programs and facilities for the 1980's. The Committee considered the contributions that each proposed program could make in response to the scientific opportunities identified. The final priorities reflect primarily the scientific importance of each program, which is discussed in Chapter 5 (in the case of the Prerequisites for the New Research Initiatives) or in Chapter 6 (in the case of New Programs). In addition, however, the Committee took into account the technological readiness of each program, its ability to complement other programs (either present, planned, or recommended in this report), and its cost. The recommendations identify those activities and facilities that the Committee believes will yield the maximum scientific return for the resources invested.

The Committee identified a number of Prerequisites for New Research Initiatives, components of the general astronomical research enterprise that are essential to the success of major programs but that are inexpensive in comparison. Each of the Prerequisites should receive a substantial augmentation of funding.

The proposed New Programs vary widely in cost. As it is difficult to compare projects of widely differing costs, the Committee grouped the New Programs into three categories—major, moderate, and small, depending on the fraction of agency resources estimated to be required—and compared projects only with others in the same category.

Assignment of priorities to the four major new programs required intensive discussion, which resulted in unanimous agreement. Despite its relatively high cost, an instrument having the capability of the Advanced X-Ray Astrophysics Facility (AXAF) is the highest priority for the 1980's. With the results furnished by the *Einstein* (HEAO-2) Observatory, x-ray astronomy became a vital part of astronomical research across a broad front. AXAF will provide astronomers with the capability to carry out a new generation of x-ray observations for a decade or more into the future, complementing the powerful capability of ST in the optical-ultraviolet region and the capability of the VLA and the VLB Array in the radio region.

Both the VLB Array (priority 2) and the New Technology Telescope (NTT, priority 3) are extremely important ground-based facilities, but for different reasons. The VLB Array will be able to make observations with extraordinarily high angular resolution of the enigmatic central regions of quasars and will make it possible to determine the size of our Galaxy by entirely new methods. NTT, on the other hand, will make measurements of a more conventional type in the optical-infrared region but with a major advance in speed and penetrating power by comparison with telescopes now available. There is a pressing need to pursue the implications of data from other wavelength regions, such as radio and x-ray observations, with optical and infrared spectra of planets, stars, galaxies, and quasars. With double the collecting area now provided by all 20 of the largest telescopes in the world, NTT will greatly accelerate such work and will also make it possible to study spectra of objects whose great distance has so far made it impossible to do so.

In the view of the Committee, the construction of the VLB Array and NTT are both of high priority. It will be necessary to complete further, extensive design studies before NTT is initiated, however, whereas the VLB Array utilizes proven technology. For this reason it is recommended that the construction of the Array proceed immediately on completion of the overall Array design, while the construction of NTT should begin as soon as the more substantial design and cost-estimate studies required for this facility have been completed.

The Committee believes that a Large Deployable Reflector (LDR) in space would provide a powerful new capability for sensitive spectroscopy and accurate imaging at wavelengths between the far-infrared and millimeter regions of the electromagnetic spectrum. The

LDR represents the culmination of a coherent, evolutionary program of infrared astronomy proceeding from the capabilities of the IRAS Explorer satellite, the Shuttle Infrared Telescope Facility, and NTT. The priority assigned by the Committee (number 4) reflects the fact that the technology to build such a facility in space is not yet mature; however, it could become so in the second half of the 1980's.

The major new programs taken as a whole—together with the completed VLA, ST (to be launched in 1985), and Gamma Ray Observatory (projected for launch in 1987–1988)—will permit impressive advances to be recorded in astronomical capability across the entire electromagnetic spectrum, from gamma rays to radio waves, including x-ray, ultraviolet, optical, infrared, and submillimeter radiation. The capability to observe individual objects at various wavelengths greatly increases the scientific value of the data obtained.

The recommended moderate new programs address a variety of opportunities in astronomy. One of them, an astronomical Search for Extraterrestrial Intelligence (SETI), was given special consideration by the Committee following the presentation of the report of the Working Group requested to study the possibilities in this area. While the Committee recognized that this endeavor has a character different from that normally associated with astronomical research, intelligent organisms are as much a part of the Universe as stars and galaxies; investigating whether some of the electromagnetic radiation now arriving at Earth was generated by intelligent beings in space may thus be considered a legitimate part of astronomy. Moreover, the techniques that can now be most effectively brought to bear on a SETI program for the 1980's are those of astronomy. In considering implementation of the moderate programs, the Committee did not believe it necessary to assign priorities as precisely as it did for the major projects.

The four small new programs recommended were selected from a much larger list of small-scale projects that came to the attention of the Committee. Although there is no implication that other programs of this scale should not also receive serious consideration, the Committee believes that the programs listed are particularly noteworthy. In connection with the last of these—a temporary program to maintain scientific expertise at U.S. universities—the Committee benefited from the report of the National Research Council's Commission on Human Resources (*Research Excellence Through the Year 2000*, National Academy of Sciences, Washington, D.C., 1979), which recommends a more broadly based program of support to young faculty members engaged in scientific research generally. While sympathetic to the

Commission's findings, the Committee believed it more appropriate in the present context to recommend a program restricted to astronomical research activity, to be funded by the Astronomy Division of NSF.

Decisions as to whether to pursue large, medium, or small-scale projects within a given time frame should be made on the basis of current availability of funds, in consultation with the appropriate advisory committees. At the same time, it is recognized that the available funding may not reach the level requested, in which case some modifications may be required. It is recommended that any required stretch-outs or changes in priorities be carried out in such a way as to maximize the scientific return of the overall program. This requires a careful assessment of various factors, including the personnel available in the relevant subdisciplines of astronomy, the balance of effort among subdisciplines, and the need for complementary observations from different wavelength regions and of different targets. Such assessments are best carried out by the appropriate advisory committees on a continuing basis.

SUPPLEMENTARY TABULATION OF PROGRAM CHARACTERISTICS

Table 2.2 presents the status, specifications, and scientific objectives of a selection of the programs discussed earlier in this chapter. It should be emphasized that this listing is intended to supplement such discussions, rather than to provide summaries of all of them. Such a listing will facilitate comparisons among most of the major programs discussed earlier.

TABLE 2.2 Supplementary Tabulation of Program Characteristics

Program	1982 Status	Specifications	Problems Addressed
Implementation of Major Programs Recommended by Greenstein Report			
Very Large Array (VLA) Radio Telescope	Construction complete; operated by National Radio Astronomy Observatory	Radio wavelengths, 1 cm to 21 cm Twenty-seven 25-m antennas Total diameter, 35 km	High-resolution studies of radio sources; radio jets of galaxies and quasars; compact H II regions; planets and solar features
High Energy Astronomical Observatories (HEAO's)	HEAO-1 (1977–79) HEAO-2 (*Einstein*) (1978–81) HEAO-3 (1979–81)	X-ray, gamma-ray Imaging and spectroscopy of high-energy sources by *Einstein* (HEAO-2)	Nature and statistics of high-energy sources; normal stars as x-ray sources; hot gas in clusters of galaxies; quasars and active galaxies; nature of x-ray and gamma-ray background
Approved, Continuing, and Previously Recommended Programs			
Space Telescope (ST)	Shuttle launch, 1985	Ultraviolet, optical, near-infrared, 1200–12,000 Å Aperture 2.4 m $10\times$ ground-based angular resolution (0″05 at 5000 Å) Limiting magnitude, 28	Stellar populations of Milky Way Galaxy; individual stars in nearby galaxies; quasars; details of planetary surfaces and galaxies; spectra of hot stars and interstellar gas
Gamma Ray Observatory (GRO)	Approved; started by NASA in FY 1981–82; projected Shuttle launch 1987–88	Gamma-ray High angular resolution High spectral resolution	Nucleosynthesis in supernova explosions; nature of gamma-ray burst sources; pulsars; active galaxies; cosmic-ray interactions in interstellar medium; gamma-ray background

Solar Optical Telescope (SOT)	Selected by NASA as first major astrophysics facility for Spacelab; projected launch 1987–88	Ultraviolet, optical, near-infrared 10× to 50× ground-based angular resolution (0″.1 at 5000 Å, 0″.02 at 1100 Å)	Dynamics of the solar photosphere, chromosphere and corona; solar magnetic field
Shuttle Infrared Telescope Facility (SIRTF)	Proposed NASA Spacelab facility recommended by CSAA	Infrared, submillimeter, 2–300-μm wavelength Wide field 10^3× sensitivity of present telescopes	Statistical study of infrared sources; molecular clouds and star formation; evolution of distant galaxies; planetary atmospheres
25-Meter Millimeter Wave Radio Telescope	Recommended by Greenstein report; next major ground-based facility projected by NSF long-range plan for astronomy	1–10-mm wavelengths Aperture, 25 m Located at high dry site	Interstellar molecular clouds and star formation; mass loss from stars; intergalactic gas in clusters of galaxies; interstellar chemistry

Major New Programs Recommended by Astronomy Survey Committee

Advanced X-Ray Astrophysics Facility (AXAF)	Proposed Shuttle-launched, free-flying observatory	X-ray 100× sensitivity and 10× angular resolution of *Einstein* (HEAO-2) Capability for polarimetry High spectroscopic sensitivity and resolution	Stellar x-ray sources throughout Galaxy; x-ray sources in nearby galaxies; distant active galaxies and quasars; clusters of galaxies; nature of x-ray background radiation; supernova remnants
Very-Long-Baseline (VLB) Array	Proposed ground-based facility	Radio wavelengths 100× angular resolution of any image-forming telescope	Compact objects in molecular clouds; center of the Galaxy; nuclei of galaxies; quasars; testing General Theory of Relativity; precise distance scale in Galaxy; geodesy and space navigation

TABLE 2.2 *Continued*

Program	1982 Status	Specifications	Problems Addressed
New Technology Telescope (NTT)	Proposed ground-based facility	Optical, infrared (0.3–20-μm wavelength) High, dry site High angular resolution at 20 μm Aperture 15 m; 9× area of Palomar 5-m reflector	Spectroscopy of faint sources; chemical composition and evolution of galaxies; red shifts of distant galaxies and quasars; very high spectral resolution of stars and planets; composition of intergalactic gas; infrared spectra of molecular clouds and protostars
Large Deployable Reflector (LDR) in space	Proposed space facility	Far-infrared, submillimeter Aperture, 10 m Arcsec resolution at 20-μm wavelength	Star formation in Milky Way and other galaxies; active nuclei of galaxies; molecular lines in planetary atmospheres and interstellar clouds
Moderate New Programs Recommended by Astronomy Survey Committee			
Augmentation to Explorer Satellite Program	Proposed to enhance effectiveness of NASA Explorer Program	Variety of missions Wide range of wavelengths Cost-effectiveness	Spectroscopy of x-ray sources; isotopic composition of cosmic rays; soft x-ray studies of a variety of very hot objects; high-energy transient phenomena; corona and interior of the Sun
Far-Ultraviolet Spectrograph in space	Under study by NASA; interest by European astronomers	1-m telescope High-resolution spectroscopy at wavelengths below 1200 Å Reach 12th magnitude at high resolution, 17th magnitude at low resolution	Key atoms and molecules in interstellar medium and planetary atmospheres; galactic halos; extended atmospheres of hot stars; chromospheres and coronas of all classes of stars

Space Very-Long-Baseline (VLB) Interferometry Antenna	Under study by NASA	Extensions of ground-based VLBI; 3 × angular resolution of ground-based VLB Array, and cleaner beam. More complete sky coverage	Compact objects in molecular clouds; center of Galaxy; nuclei of galaxies; quasars; testing General Theory of Relativity; precise distance scale in Galaxy
Optical/Infrared Telescopes in the 2–5-m class	Proposed ground-based telescopes	Special consideration to projects obtaining significant nonfederal funding. Take advantage of radical reductions in cost of telescopes built with new technology	Transient phenomena; long-term surveys; follow-up of observations in other wavelength regions, particularly those from space; development of focal-plane instruments
Advanced Solar Observatory (ASO) in space	Proposed Shuttle-launched, free-flying facility; incorporates Solar Optical Telescope	Simultaneous observations at gamma-ray, x-ray, extreme-ultraviolet, and optical wavelengths. High angular resolution (better than 1 arcsec at all wavelengths)	Structure of solar core; dynamics of the Sun; solar flares; solar wind; heating of chromosphere and corona
Cosmic-ray experiments	Proposed Shuttle-launched, free-flying instruments	Long duration (≥6 mo). Space platform	Origin of elements in stars; nature of cosmic particle accelerators; interstellar gas; violent cosmic processes
Astronomical Search for Extraterrestrial Intelligence (SETI)	Proposed new program	Long-term effort at modest level. Start with search of radio wavelengths	Possible detection of extraterrestrial intelligence; parameters of extrasolar planetary systems

TABLE 2.2 *Continued*

Small New Programs Recommended by Astronomy Survey Committee

Program	1982 Status	Specifications	Problems Addressed
10-m Submillimeter-Wave Antenna	Proposed new program	Submillimeter wavelengths High, dry site 10-m aperture High angular resolution, up to 8″5	Higher rotational transitions of molecules in interstellar clouds; star formation; interstellar chemistry; mapping neutral atomic carbon in Galaxy
Spatial Interferometer for Mid-Infrared	Proposed new program	Mid-infrared wavelengths (2–20 μm, optimized at 10 μm)	Circumstellar dust and stellar mass loss; star formation; nuclei of active galaxies; astrometry; angular diameters of stars
High-Precision Optical Astrometry	Proposed new program	Innovative astrometric devices for relative positions with accuracy of 10^{-4} arcsec Possible construction of astrometric telescope	Detection of Jupiterlike planets around other stars; luminosities, radii, and masses of stars from improved stellar distances to 1 kiloparsec; improve precision of extragalactic distance scale
Temporary Program to Maintain Scientific Expertise at U.S. Universities	Proposed new program	Match one-half salary provided by universities for positions at assistant-professor level 10 to 20 five-year positions Open national competition	Assures flow of younger researchers into astronomy faculties

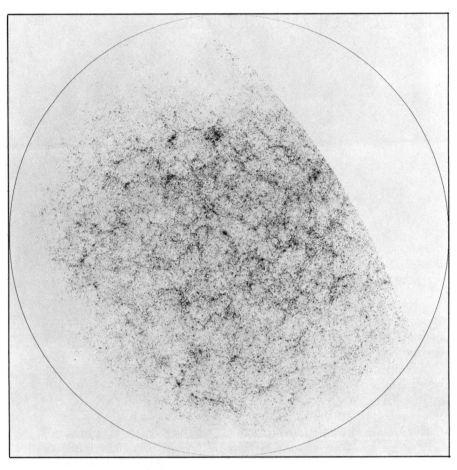

Angular distribution of galaxies brighter than blue magnitude 19, from the Lick Catalog of Galaxies. (Photo courtesy of M. Seldner, B. Siebers, E. J. Groth, and P. J. E. Peebles, Princeton University)

3

Frontiers of Astrophysics

Probes of Large-Scale Structure

As seen on photographs taken with the largest telescopes, galaxies appear to drift in the depths of space like motes in a sunbeam. Everywhere they are clumped in groups containing a few galaxies, and occasionally in clusters of a thousand or more. Some clusters clump in superclusters 50 megaparsecs or more across. On even larger size scales, however, groups and clusters of galaxies seem to be distributed nearly at random, the number in a given volume of space being about the same throughout the Universe.

This uniformity of the distribution of matter on very large scales invites comparison between the observations and a simple model of the Universe, or cosmology, derived for a uniform distribution of matter from Einstein's General Theory of Relativity. According to this model, the geometry of space–time is curved by matter, and the curvature forces the matter to move: at any epoch the Universe must be either expanding or contracting. Hubble's discovery in 1929 that the Universe is actually expanding forces us to confront a bizarre implication of the theory: that an expanding Universe must have originated in a powerful explosion—referred to as the big bang—before which neither time nor space had any meaning.

In the half-century since Hubble's momentous discovery, astron-

37

omers have been probing the Universe in space and in time. Using radio and optical telescopes, they have found objects so distant that they are receding from us at 90 percent of the speed of light. With microwave antennas, they have discovered a faint radio noise that they interpret as the remnant of the big bang itself. From the theory of the nuclear reactions that must have taken place during the first 3 minutes, they have calculated the abundances of key elements and isotopes such as hydrogen, deuterium, and helium, which were produced in the big bang; with ground-based telescopes and ultraviolet spectrographs in Earth orbit they have verified that the actual relative numbers of these atoms in space agree surprisingly well with theoretical predictions.

The big bang has become the standard model with which to compare observations. This is not to say that it is completely correct: the data are imprecise; their interpretation may be in error; and the theory could be wrong. A central problem for the future is the further development of the big-bang model and its testing against all available observations.

The big-bang model requires that matter is distributed uniformly on large scales. By using a variety of approaches, it is now possible to test whether this is true. Observers have used apparent magnitudes as a rough measure of the distances of galaxies; plotting the directions of galaxies in various distance ranges, they have found that on scales exceeding 100 megaparsecs, galaxies are distributed rather uniformly. One can obtain the precise location of each galaxy in three dimensions by determining its red shift spectroscopically. Recording the spectrum photographically is time consuming, but the recent development of electronic array detectors has speeded up the recording of spectra so greatly that red-shift surveys of thousands of galaxies are now possible. The resulting three-dimensional distribution appears to be uniform on the largest scales. It is anticipated that red-shift surveys of much more distant galaxies will be completed during the 1980's.

X-ray and gamma-ray astronomy also tell us about the large-scale distribution of matter. A diffuse background emission not attributable to known sources appears in both spectral regions; its near isotropy proves that it cannot originate within the Galaxy but must instead originate at distances comparable with the size of the Universe itself. The High-Energy Astronomical Observatory-1 (HEAO-1) x-ray observatory established that the x-ray background is highly isotropic and that its spectrum between a few and about 60 kiloelectron volts (keV) agrees closely with the radiation expected from a gas having a temperature of about 500 million degrees, leading to the suggestion that such gas is distributed uniformly between the galaxies. The *Einstein* (HEAO-2) x-ray observatory, on the other hand, discovered that individual quasars at large distances are powerful x-ray sources in the few-keV range—powerful enough, in fact, that quasars at even larger distances than can be detected individually by the *Einstein* x-ray observatory must account for a substantial fraction of the observed x-ray background in the few-keV range. As some quasars have also been found to be powerful gamma-ray sources, the gamma-ray background may also be due to quasars. It is still not clear, however, how quasar spectra would sum up so as to mimic the spectrum of hot gas. The Advanced X-Ray Astrophysics Facility (AXAF) recommended in this report can observe sources 100 times fainter than could the *Einstein* x-ray observatory and can thus determine whether faint quasars account for the observed background at energies of a few keV. Measurements of faint quasars by the Gamma Ray Observatory (GRO) will give similar information for the background at gamma-ray energies. If it proves that the x-ray and/or gamma-ray backgrounds are actually due to quasars, the fact that the background is highly isotropic requires that matter at great distances is distributed very uniformly. If, on the other hand, intergalactic gas is responsible for at least part of the x-ray background, one can infer that it is distributed uniformly; moreover, the amount of gas required is an important datum for the theory of evolution of galaxies.

The cosmic microwave background radiation also gives information about the large-scale structure of the Universe. Precise measurements have revealed a smooth variation in its intensity over the sky that is attributable to the Earth's motion through the cosmos. The observed variation is unexpectedly large, corresponding to a velocity of 500 km/sec for the Local Group of galaxies with respect to distant matter. The same measurements reveal no other certain variations larger than 0.03 percent, indicating that the Universe was highly uniform at the time the background radiation last interacted with matter. Ground-based experiments indicate that the spectrum of the microwave background radiation does not deviate significantly from thermal, as predicted by the big-bang model, but a balloonborne submillimeter experiment points to discrepancies that are difficult to explain. Both variations in intensity with direction and deviations from a thermal spectrum will be measured over the entire spectral range with improved precision (about 0.01 percent) by the Cosmic Background Explorer (COBE) mission planned by NASA.

Expansion Time Scale

The big-bang model predicts that galaxies should move away from each other with velocities that are proportional to their separations. Slipher and Hubble found evidence for such a relationship in extensive measurements of the brightnesses and red shifts of galaxies during the 1920's; the constant of proportionality between a galaxy's velocity and its distance is called the Hubble constant. According to relativistic models, the reciprocal of the Hubble constant (the "Hubble time") is roughly equal to the present age of the Universe—that is, the time that has elapsed since the big bang.

Determining the value of the Hubble time requires the measurement of the distances of remote galaxies, using a "ladder" of interconnected distance scales determined by different methods; each step of the ladder reaches further into space. Hubble's own estimate for it was 2 billion years. It has since been revised several times—to 5, then to 10, and then to 20 billion years; the latest estimates are between 10 billion and 20 billion years. Each revision has been the result of a major advance in understanding the properties of stars or galaxies that are used to construct the ladder of distance scales.

The value of the Hubble time enters all cosmological calculations in a fundamental way. To find its true value, each step of the ladder of distance scales must be secure, and any contributions to the velocities of galaxies that are not due to the expansion of the Universe must be taken into account. An example of the latter effect is the motion of the Local Group of galaxies revealed by study of the cosmic background radiation; when this is taken into account, a more consistent set of data for the Hubble time emerges.

Refinement of the distance ladder will take much more work. Development of more precise astrometric methods, as recommended in this report, will make possible a more accurate measurement of the distance to the Hyades star cluster, the first step in the ladder of cosmic distance scales. Because of its extremely faint limiting magnitude, Space Telescope (ST) will for the first time resolve Cepheid variable stars in the Virgo cluster, thereby eliminating an uncertain intermediate step of the distance ladder. The continued deployment of advanced optical detectors at ground-based telescopes will make possible the rapid measurement of red shifts of galaxies at moderately large distances, where the velocity field should be one of nearly pure expansion; ST can determine the distances of the same galaxies by comparing the brightness of their globular clusters with the brightness of those in the galaxies of the Virgo cluster, whose distances are known accurately.

The Early Universe

The cosmic microwave background radiation carries information about the Universe before it was about 1/100,000 of its present age, so the COBE experiment is fundamental to studies of the early Universe. Other clues depend on the nucleosynthesis of various elements and isotopes in the first 3 minutes. Theoretical predictions of their abundances depend critically on the amount of ordinary matter present during that period. If the amount is low, the resulting deuterium abundance would be high and the helium abundance low; if the amount is high, the opposite would be the case. Present information on the abundance of deuterium and helium in interstellar space in our Galaxy, taken at face value, indicates that the amount of matter is too low by a factor of 10 for its gravitation to be able to halt the expansion of the Universe.

However, helium has been produced and deuterium has been destroyed in stars, so present abundances in the Galaxy may not be the same as in the primordial gas that emerged from the big bang. Abundances in intergalactic gas, if it exists, should be primordial. Astronomers have discovered absorption-line systems in distant quasars that probably originate either in clouds formed by the outward ejection of thick shells of gas from the quasar itself or in intergalactic clouds lying along the line of sight. In the first case, the phenomenon would resemble the late stages occurring in the stellar outbursts known as novae. In the second case, the clouds should contain very little carbon or other medium-weight elements, which are telltale signs of stellar nucleosynthesis, because such gas would never have been inside a galaxy. The gas in such clouds would be a good candidate for the study of primordial helium and deuterium. Observations of helium and deuterium in such gas, however, must be made at much shorter wavelengths than are accessible to ground-based observatories; they require ST. With ST we can study helium lines in clouds of red shift greater than unity and deuterium lines in clouds of all but very low red shifts.

Groups, Clusters, and Superclusters

The grouping of galaxies on various size scales can be studied by calculating the statistical correlations between the observed positions of galaxies. In earlier studies, the apparent magnitudes of galaxies were taken as measures of their distances, and their positions then follow from their observed directions in the sky. The calculated correlations between the positions derived in this way decrease as an inverse power of the distances between pairs of galaxies. A simple model to explain this is based on gravitational clustering of point masses, which are initially distributed at random but which then move under their mutual gravitation as the Universe expands. This model reproduces many of the features of the observed clustering of galaxies, so that galaxies may have formed early in the expansion of the Universe and clumped together later by gravitation.

Recent observational work, however, has brought out an unexpected new feature in the distribution of galaxies. Aided by redshift measurements, which furnish the distances of galaxies much more accurately than estimates based on their apparent magnitudes, astronomers have found that groups of galaxies outside of clusters are not sprinkled at random through space but instead lie in great sheets between the clusters, leaving vast empty regions between. To explain this may require a new theoretical model, in which galaxies formed rather late. At first, giant turbulent cells of gas collided, compressing the gas into sheets; only after the sheets formed did the galaxies condense from them and then begin to clump together as in the earlier model.

Two kinds of data are required if we are to understand the formation and clumping of galaxies. First, red-shift surveys embracing a large number of galaxies are needed. For the nearer galaxies, it is feasible to obtain red shifts with currently available telescopes of moderate size, equipped with array detectors and fast spectrographs. To penetrate more deeply into space, however, large telescopes will be needed. Telescopes of the 5-m class will make important contributions, but only a new telescope of the 15-m class, such as the New Technology Telescope (NTT), can measure the red shifts of galaxies at large distances rapidly enough to accumulate the required number of galaxies. The raw speed of NTT, made possible by its order-of-magnitude increase in collecting area over the previous largest telescopes, is critical for this project.

Hidden Mass and the Fate of the Universe

For the past 20 years, astronomers have been increasingly puzzled by the "hidden mass" problem: the matter that constitutes most of the mass of the Universe is invisible. The spectra of galaxies indicate that, like our own Milky Way Galaxy, they contain normal stars; however, the internal motions in galaxies are so large that they would fly apart if the only gravitational attraction holding them together were that of the stars we see. There must be additional mass present in some form that is hidden from our immediate view—enough to supply the gravitational attaction required for stability. The rotational velocities observed in spiral galaxies demonstrate that the amount of hidden mass inside a given radius increases approximately linearly with radius out to distances of nearly 100 kiloparsecs. Similar results emerge from studies of groups of two or more galaxies: their masses must be at least 10 times greater than the masses of all the visible stars in them.

Solution of the hidden-mass puzzle is a major goal of astronomy in the decade ahead. The first task is to find how it is distributed. The velocities of globular clusters in the outer reaches of galaxies reflect the strength of the local gravitational field and hence the distribution of mass in the parent galaxy. Since globular clusters in even relatively nearby galaxies are extremely faint, spectroscopic measurements of their velocities can be made only with a telescope as large as NTT. Galaxies themselves can serve as probes of the distribution of mass in clusters and superclusters of galaxies. Since galaxies are much brighter than globular clusters, work on clusters of galaxies is already proceeding with intermediate-sized telescopes. However, measurements of velocities of galaxies in distant clusters are essential to determine how the distribution of mass has changed with time; this will require observations with NTT.

Various possibilities have been suggested to account for hidden mass: diffuse gas, massive neutrinos, collapsed stars (white dwarfs, neutron stars, black holes), and faint red dwarfs.

Diffuse gas can be ruled out as a dominant component of either galaxies or clusters of galaxies through radio, optical, and x-ray observations; although 100-million-degree gas exists in clusters of galaxies, the amounts are not sufficient to hold the clusters together. Massive neutrinos, if they exist, might fall into clusters of galaxies, and possibly even into galaxies themselves, thus contributing to the hidden mass.

Collapsed stars of various types could in principle constitute much of the hidden mass; however, such stars are the descendents of massive main-sequence stars and so would dominate the total mass only if, at early epochs of star formation, massive stars dominated the total mass of main-sequence stars. Just the contrary is observed to be the case for star formation in our Galaxy near the Sun: faint red dwarfs, which are of low mass, are so numerous that they account for most of the mass bound up in stars. One could speculate that there were many more massive stars in the outer parts of galaxies during the early stages of galaxy evolution, so that large numbers of collapsed stars would exist there today. However, if that were so, one would expect a higher concentration of heavy elements in the outer parts of the galaxies, since massive stars synthesize heavy elements and eject them into the interstellar medium; this is contrary to observation.

Faint red dwarfs could also account for the hidden mass, as large numbers of them in the outer parts of galaxies would be consistent with both the lower concentrations of heavy elements and the lower light levels observed there. It may just prove possible to test this hypothesis by using the recent discovery that red dwarfs are relatively luminous sources of coronal x rays. AXAF will be able to detect such red dwarfs by observing their integrated coronal x-ray emission if they are numerous enough.

The hidden-mass problem is intimately connected with the question of the ultimate fate of the Universe. According to the big-bang model, the Universe will continue to expand forever if the amount of matter in it is less than a critical value calculated to be between 0.5×10^{-29} and 2×10^{-29} g in each cubic centimeter. If the amount of matter exceeds the critical value, the present expansion will reverse at some time in the distant future, and the Universe will collapse back into a singular state similar to the big bang. The observations of deuterium and helium discussed earlier suggest that the amount of ordinary matter is only 10 percent of the critical value, so that only massive neutrinos could raise it above the critical value. A lower limit on the total amount of mass in all forms is obtained from the masses of clumps in the distribution of galaxies; current estimates suggest that the aggregate amount of matter in such clumps may be as much as 40 percent of the critical value. Since this is larger than the upper limit on the amount of ordinary matter obtained from observations of helium and deuterium, massive neutrinos may conceivably account for most of the matter in the Universe. Massive neutrinos are discussed further in the last section of this chapter.

EVOLUTION OF GALAXIES

The Study of Galaxies

Like the Galaxy in which we live, the 100 billion or more galaxies in the visible Universe are fascinating systems in their own right. As the nuclear and gravitational energy stored in them is released, it is likely that galaxies evolve toward objects evermore structured and compact.

Among the variety of forms that galaxies take, Hubble discerned several recurrent patterns—spirals, ellipticals, lenticulars, and irregulars; these patterns have still not been completely explained theoretically. Ellipticals and lenticulars are nearly devoid of interstellar gas and dust, while spirals and irregulars contain gas and dust, as well as young stars formed recently from them. Until recently, the gas and dust in spiral galaxies other than our own could be studied with high angular resolution only at optical wavelengths, by imaging the dark interstellar dust clouds and the luminous gas clouds heated by bright young stars. Now the Very Large Array (VLA) radio telescope can image galaxies both in the 21-cm line produced by interstellar atomic hydrogen and in the synchrotron radiation produced by relativistic electrons gyrating in interstellar magnetic fields; it can thus trace the distribution and state of the interstellar medium with angular resolution comparable with that of optical telescopes.

As in all fields of astronomy, spectroscopy is the key to deeper understanding. Ground-based optical spectroscopy of galaxies demonstrates that a major component of most galaxies is stars of various masses and ages, like those in our Galaxy. However, present ground-based telescopes are hard pressed to obtain the spectra of extremely faint subsystems of galaxies, such as individual giant stars, regions of ionized gas, and globular clusters; they are too small to permit collection of photons at a sufficiently high rate. NTT, with its order-of-magnitude increase in collecting area, can obtain the spectra of such objects, thus making possible a whole new range of studies related to chemical composition, distribution of stellar masses, and rotational and random velocities within galaxies. For a galaxy of a given red shift, NTT will make possible studies with much higher spectral resolution; for the same spectral resolution, it can carry out studies on galaxies of much higher red shift. The latter capability is crucial for analysis of objects of large red shift that will be discovered by ST.

One of the most striking capabilities of the new instruments recommended for the 1980's is the systematic exploration of the dependence of various galactic properties on red shift at greater and greater cosmological distances. Big-bang models of the Universe predict such a dependence because the evolution of galaxies with time translates into changes with lookback time, and hence with red shift. ST and NTT will be able for the first time to observe galaxies with red shifts substantially exceeding unity, corresponding to lookback times that are more than half of the Hubble time. ST can image such distant objects because the sharpness of its images makes them stand out against the background, and NTT can obtain their spectra because it has a much larger collecting area than present large telescopes. If the matter comprising the inner parts of galaxies has already settled into an equilibrium state within considerably less than a billion years after the big bang, the forms of galaxies would not depend sensitively on red shift out to red shifts of 10 or so. However, the evolution of stars and the conversion of interstellar gas into stars proceeds much more slowly and should be observable at much lower red shifts. The spectra of isolated elliptical galaxies should manifest subtle changes that reflect the evolution of the stars that they contain, while isolated spiral galaxies should in addition manifest the progressive depletion of interstellar matter, as well as its enrichment in heavy elements produced by supernova explosions. A major indirect effect will be the reduction in the number of short-lived massive stars as the gas required to form them is depleted. Failure to observe such basic predictions of big-bang theory would force major revisions in current thinking.

Formation of Galaxies

The first relativistic models of the big-bang Universe were derived by Friedmann in 1922. For simplicity, he assumed that matter is distributed absolutely uniformly. Although this assumption conflicts with the existence of stars and galaxies, the model is useful because matter is in fact distributed quite uniformly when averaged over large distances. Still, the origin of galaxies in a big-bang model is an unresolved problem.

Many properties of galaxies can be explained at least qualitatively if it is assumed that they originated in small fluctuations in the amount of local matter in the early Universe. At that time, the behavior of matter was governed by the pressure exerted by the cosmic background radiation. Two types of density fluctuations could have existed. One type, so-called isothermal fluctuations, would have led to gravitationally unstable clumps of matter if they had involved more than 10^5 to 10^6 solar masses; another type, adiabatic fluctuations, would have led to gravitationally unstable clumps if they had involved more than 10^{13} to 10^{14} solar masses. In both cases, instability would have set in about 100,000 years after the big bang, and as a result, the matter in the fluctuations would soon cease to participate in the cosmic expansion, would then become more dense as self-gravitation drew the gas together, and would ultimately form discrete gas clouds of various masses.

The Cosmic Background Explorer (COBE) satellite will yield important information on the proposed instability process by observing the disturbances in the background radiation that would accompany any density fluctuations in the early Universe. Adiabatic fluctuations, which involve variations in temperature and hence in the intensity of the cosmic background radiation, would result in intensity variations on angular scales of a few degrees if the masses involved in the fluctuations are about those of clusters of galaxies. Complementary information about fluctuations on the smaller angular scales corresponding to individual galaxies (less than a degree) will be obtained by the Large Deployable Reflector (LDR) in space. The theory of adiabatic fluctuations has been worked out in detail for the case in which there is a random collection of initial fluctuations of various sizes and masses. Fluctuations involving 10^{13} to 10^{14} solar masses, usually identified with groups and clusters of galaxies, should form clouds first; the formation of galaxies would have taken place later within these clusters and groups. Clusters and superclusters containing more than 10^{13} to 10^{14} solar masses must have formed through later gravitational clustering of the original mass aggregations of this size.

Alternatively, isothermal fluctuations may have dominated the initial stages of galaxy formation. In this case, the first objects to form must have had masses from 10^5 to 10^6 solar masses, and galaxies must have been built up later by gravitational clustering of these smaller objects. The fact that globular clusters containing 10^5 to 10^6 solar masses are so common would be a natural result of isothermal fluctuations. If galaxies formed out of objects having 10^5 to 10^6 solar masses, then groups and clusters of galaxies must have formed subsequently through gravitational clustering of the galaxies themselves. This process can be modeled with computers by treating each galaxy as a point mass and calculating its gravitational interactions with its neighbors. Extensive simulations of gravitational clustering have been carried out in this way during the past decade; the results agree with observations in some respects, but they do not predict the large holes devoid of galaxies that have been observed between clusters. It is still uncertain whether galaxies or clusters of galaxies originated first.

None of the existing computer simulations of either galaxy collapse or clustering addresses the origin of the fluctuations themselves. Current attempts to answer this important question, based on Grand Unified Theories of elementary particles, are encouraging.

The collapse of a gas cloud to form a disk galaxy like the Milky Way and the subsequent formation of the first generation of stars are both believed to have taken place during the first billion years of galactic evolution. These processes could therefore be observed only in galaxies so remote that the radiation now received from them was emitted in their early youth. Although presumably very faint, this radiation, red shifted by a factor of 10 or more because of the cosmic expansion, might be detected by the next generation of instruments. Optical and ultraviolet radiation from young galaxies, if not absorbed by dust near the galaxy or in intragalactic space, would be red shifted into the infrared region, where the extreme sensitivity of the Shuttle Infrared Telescope Facility (SIRTF) would permit it to be detected. Overlapping shock waves from early supernova explosions could heat the interstellar gas to temperatures as high as 1 billion degrees, generating dust-penetrating x rays with energies up to 100 keV. These x rays, now red shifted into the few-kiloelectron-volt range, should be detectable by AXAF according to recent calculations. If a galaxy like ours that had just been born were discovered, it would present a breathtaking opportunity for study.

Evolution of Galaxies

Once galaxies form, they evolve slowly as dying stars inject newly synthesized atomic nuclei into the remaining interstellar gas, from which new generations of stars are formed. Current models suggest that most of the matter entering the interstellar medium is ejected by stars of moderate mass, which evolve into red giants and then planetary nebulae, as red giants lose most of their outer layers in low-velocity stellar winds, while the deeper layers are thrown off in final outbursts that produce planetary nebulae. Most of the energy injected into the interstellar medium, on the other hand, comes from the explosions of massive stars, in which an entire star is disrupted to form a supernova. Supernova explosions are also the principal sources of heavy elements.

Only stars whose mass equals or exceeds that of the Sun evolve significantly in a Hubble time, so that during the generations of stars that have occurred since the galaxies formed, much of the interstellar medium has found its way into low-mass stars that have evolved little over the lifetime of the Universe. Thus, interstellar matter has been continually enriched in heavy elements, while its mass has been continually reduced as more of it is converted into slowly evolving stars.

Star formation apparently proceeds at different rates in spirals and ellipticals, since at present spirals contain large amounts of interstellar matter, while ellipticals contain little. It has been suggested that early in the life of ellipticals, the star-formation rate was high enough that frequent supernova explosions were able to drive the remaining gas out of the galaxy in a so-called "galactic wind," thus quenching further star formation. Supernova explosions apparently occur frequently enough today to keep ellipticals swept clean of interstellar matter. In spirals, on the other hand, the initial supernova rate may not have been great enough to cause a catastrophic purging of the interstellar medium, and so a sufficient amount of interstellar matter still remains today to support active star formation. Thus, the different amounts of interstellar medium in ellipticals and spirals may be a consequence of different initial rates of formation of massive stars, which become supernovae; why the rates should have been different is an important unsolved problem of galactic evolution that will be addressed through the study of galaxies of large red shift.

The existence of galactic winds is in accord with current x-ray observations of rich clusters of galaxies, in which most of the galaxies are either ellipticals or their close cousins, the lenticulars. In many such clusters there is a hot, x-ray-emitting intergalactic medium, whose total mass and chemical composition are consistent with the supposition that it has accumulated from galactic winds. The removal of gas from galaxies is aided by the intergalactic medium, once established, because it sweeps gas from galaxies as they move through it. There is evidence for this process in the x-ray and radio images

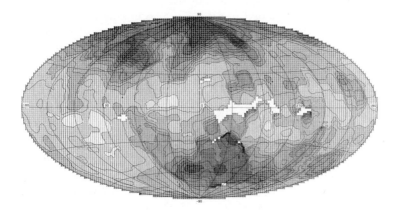

Map of the sky in soft x-rays (0.15–0.28 keV) from a series of rocket flights. (Photo courtesy of the University of Wisconsin x-ray group)

of certain galaxies that are being swept at present. One can understand the absence of gas-rich galaxies in rich clusters of galaxies along such lines, but a detailed understanding awaits x-ray observations with AXAF, which, because its sensitivity and angular resolution are greater than those of the *Einstein* x-ray observatory, can observe the process of sweeping at greater distances.

Any realistic scenario for galactic evolution must take into account the effects of cosmic rays—the relativistic particles whose presence is inferred from experiments with charged-particle detectors in space. Constituting a prominent and permanent component of our Galaxy, cosmic rays have isotopic abundances that imply that they remain tens of millions of years in the Galaxy, most of which time is spent in the Galactic halo. In effect, cosmic rays constitute a relativistic gas with a pressure comparable with that exerted both by random motions of the interstellar gas and by the interstellar magnetic field; the cosmic-ray gas therefore plays a critical role in the equilibrium inflation of the gaseous disk, in the fragmentation of the interstellar medium into molecular-cloud complexes, and, presumably, in the support and activation of halos of galaxies. Determining the origin and propagation of cosmic rays is therefore an important aspect of the overall effort to understand the course of galactic evolution. Cosmic-ray experiments on the Space Shuttle will advance these goals by supplying new information about the composition, energy spectra, and isotope distributions of the cosmic rays themselves.

Verifying our ideas of galactic evolution will require intensive studies of our own and nearby galaxies, as well as observations of galaxies so remote that their properties appear different from those of the more evolved galaxies nearby. It is now understood that stellar mass loss has a profound effect on galactic evolution. With its high sensitivity, SIRTF will discover many more cool red-giant envelopes, in which the stellar light is degraded to infrared radiation by large quantitites of embedded dust; submillimeter, millimeter, and infrared telescopes, including the LDR in space, the 10-m submillimeter-wave antenna, the 25-Meter Millimeter-Wave Radio Telescope, and NTT, will permit spectroscopic observations of the spectra of molecules in such envelopes, leading to the determination of nuclear and isotopic abundances, which are clues to the nuclear processing that has occurred in the parent stars; they will also determine the velocity and mass of the outflowing gas, crucial parameters for calculating the rate at which mass is being ejected into the interstellar medium.

Because of its exceptionally faint limiting magnitude, ST will be able to identify a much larger fraction of the low-mass stars believed to lie near the Sun than has been possible up to now, thus permitting a much more accurate assessment of the mass stored in these stars; this assessment will be aided by SIRTF, which will be much more sensitive to the extremely cool faint stars than are the infrared telescopes now available. Because of its ultraviolet sensitivity, ST is expected to find many new white dwarfs, hence improving our estimates of the number of moderate-mass stars that have already undergone evolution. NTT, whose large collecting area will make it possible to obtain spectra of various subsystems of nearby galaxies, will reveal variations in abundances expected to develop as the result of different rates of evolution at different points within galaxies; the 25-Meter Millimeter-Wave Radio Telescope will permit determination of isotopic abundances of CO and other molecules with sufficient angular resolution to detect variations in abundances across the faces of galaxies; its beamwidth corresponds to 60 parsecs at M31.

By yielding sharp images of large-red-shift galaxies, ST will permit them to be classified morphologically for the first time; this is essential in order correctly to interpret observed correlations between the red shifts and other properties of galaxies. There have been tantalizing hints that certain types of galaxies are systematically bluer than their analogs nearby at lookback times as small as 3 billion to 6 billion years, but because morphological classification of these galaxies is beyond the capability of present ground-based telescopes, the interpretation of these results is unclear. NTT will play an important role in this research by permitting us to obtain optical and infrared spectra of galaxies at large red shifts, including those discovered by ST.

Interaction of Galaxies with Their Environment

An important advance in studies of galactic evolution in the last decade has been the recognition of interactions between galaxies and neighboring galaxies and/or intergalactic gas. Exchange of mass, energy, or angular momentum with the environment can modify galactic evolution in a variety of ways. It is widely believed that isolated spiral galaxies are formed with extensive halos, in which a substantial fraction of the mass is hidden in faint stars or in some other form, such as massive neutrinos. Galaxies located in binary pairs will interact with each other's halos in various ways if they are close enough together. For example, one galaxy can strip the halo material from the other by tidal forces; the material may either be redistributed within a common envelope or be entirely lost from the binary system. Since any energy or angular momentum lost in this way must come from the binary orbit, the orbit must evolve. If one galaxy of the pair has a low mass, it may end up merging with the more massive galaxy, thus further increasing its mass; such a process is believed to account for the frequent occurrence of supermassive galaxies at the centers of rich clusters. The gravitational fields of such systems will be probed through measurements of the radial velocities of faint globular clusters with NTT, as well as through studies of the distribution of hot gas around galaxies with AXAF. There is evidence from x-ray observations of the Virgo cluster that the giant elliptical galaxy Messier 87 is accreting intergalactic gas. One model of the intense production of relativistic particles in the nucleus of M87 envisions the production of energy by the flow of matter into a massive black hole located in the galactic nucleus. Such accretion of intergalactic gas in clusters could provide a virtually unlimited source of matter to power the radio galaxies often found in clusters. Here AXAF, with its ability to observe the diffuse, hot gas with high angular resolution, increased sensitivity, and high spectral resolution, will be able to examine many radio galaxies at larger distances, for evidence of the M87 phenomenon.

The whole problem of activity in the nuclei of galaxies poses a major puzzle in galactic evolution. Both elliptical and spiral galaxies (as radio and Seyfert galaxies, respectively) display activity that is highly localized in the nucleus of the galaxy. Possible physical explanations for such activity are discussed in the next section; here we mention those aspects of galactic activity that are relevant to galactic evolution in general. Do the extreme examples of galactic-scale explosions called quasars occur within galaxies? ST, with its high angular resolution, will be able to observe the parent galaxies if they are there; the light from a parent galaxy, which is lost in the glare of the quasar when studied with present ground-based telescopes, will be distinct from the quasar image when viewed with ST. Do many quasars occur within groups of galaxies, as recent ground-based observations of some quasars suggest? Again, ST, with its faint limiting magnitude, and NTT, with its ability to acquire spectra of faint objects, will answer this question even for distant quasars, thus determining whether the occurrence of a quasar in a galaxy depends on the environment of the galaxy. Is the density of stars in active galactic nuclei as high as required to explain such activity by stellar collisions? High-resolution pictures with ST will penetrate close to the centers of Seyfert nuclei and hence help to answer this question; its angular resolution of 0.05 arcsec corresponds to 25 parsecs at the distance of the nearest Seyfert galaxies. Is there an inward flow of hot gas in active ellipticals, as required by theories based on accretion by a massive black hole? The high-resolution images to be obtained by AXAF and ST, and studies of the time variability of high-energy emission to be carried out by AXAF, may help to answer this question as well.

The instruments of the 1980's will have a major impact on the study of galactic evolution. Over the entire range of phenomena—from the earliest development of fluctuations in the Universe, through star formation in collapsing young galaxies, to the slow conversion of interstellar gas to stars and the emergence of active galactic nuclei—observations by the new instruments in the gamma-ray, x-ray, ultraviolet, optical, infrared, and radio regions of the spectrum will provide important new information.

VIOLENT EVENTS

Cosmic Rays, Supernovae, and Pulsars

The Earth is constantly bombarded by cosmic rays, charged sub-atomic particles from space moving at relativistic speeds and with energies up to 10^{20} electron volts (eV). The discovery of cosmic rays early in this century provided the first hints that the Universe is not a quiescent collection of stars and planets but, rather, the scene of violent events in which particles are accelerated to relativistic energies. Until the 1950's, studies of cosmic rays were limited to those particles reaching the Earth; the bending of the trajectories of cosmic rays by interstellar magnetic fields precludes the identification of their sources by observing the directions from which they arrive at the Earth.

A breakthrough in our understanding of the role of high-energy particles in cosmic processes occurred in the 1950's, when optical and radio astronomers discovered polarized emission from supernova remnants and showed that it is synchrotron radiation from relativistic electrons accelerated and trapped in magnetic fields. Shortly thereafter, electrons were discovered in cosmic rays arriving at the Earth, in numbers agreeing with those required to explain as synchrotron radiation the nonthermal radio emission observed from the Milky Way.

We now know that the acceleration of atomic nuclei and electrons to relativistic energies occurs in astronomical systems ranging in scale from the Sun and planets to giant radio galaxies. Particle acceleration results from motion of the magnetic fields embedded in the ionized gases, or plasmas, that pervade interplanetary and interstellar space. *In situ* study of plasmas in the Earth's magnetosphere and in the interplanetary medium has shown that three processes accelerate charged particles: the passage of shock waves, the reconnection of magnetic fields of opposite polarity, and the compression of magnetic fields.

Solar physicists have observed that magnetic fields emerging from the solar surface are occasionally forced into unstable configurations in which large amounts of magnetic energy are stored. Rapid reconnection of the magnetic lines of force to form new configurations releases magnetic energy, much of which goes into accelerating large numbers of particles in a short time; the resulting high-energy particles cause solar flares. Recently, the International Ultraviolet Explorer (IUE) and the *Einstein* x-ray observatory have detected similar flare activity on other stars. Magnetic activity appears to be commonplace, and in some stars it produces flares far more energetic than those on the Sun.

Particle acceleration on a much larger scale accompanies the supernova explosions that mark the deaths of massive stars. Optical, radio, and x-ray observations of supernova remnants strongly suggest that most Galactic cosmic rays are initially accelerated in supernova explosions. Supernovae are also believed to be responsible for the synthesis of heavy elements, and thus they play a critical role in the chemical evolution of galaxies and, ultimately, in the origin of stars, of planets, and of life.

There are at least two distinct types of supernovae. Type I supernovae, which may occur in low-mass binary star systems, are discussed in the next section. Type II supernovae, which occur in massive stars located in the disks of spiral galaxies, appear to be the natural consequence of the evolution of the cores of isolated massive stars. In successive stages of core contraction and heating, thermonuclear reactions produce heavier and heavier elements, until a core of about 1.4 solar masses of iron is accumulated. If an iron core contracts, because there are no further nuclear reactions to furnish energy that would stabilize it, it quickly goes into free-fall when the internal pressure drops as a consequence of the capture of electrons by atomic nuclei and the partial photodisintegration of iron-group nuclei into free nucleons and alpha-particles. If the core is sufficiently massive, no force is strong enough to reverse its huge inward momentum, and the matter collapses to a point, forming a black hole.

For cores of lower mass, as the central density approaches and even surpasses that of a typical atomic nucleus, the collapse may be halted by the repulsive component of the strong nuclear force. What happens then is still uncertain. One possibility is that the rebound of the collapsing core arising from this sudden "stiffening" of its inner regions initiates a shock wave that propagates out of the core and into the loosely attached envelope of the star, where it heats the material to such high temperatures that a sudden surge of reactions occurs among the nuclei present. The elements heavier than iron thus created, together with other heavy elements produced by earlier nuclear burning nearer the surface of the star, are ejected by the shock wave into the interstellar medium.

The process of core collapse releases about 10^{53} ergs, an amount of energy released if a star having a tenth of the mass of the Sun were completely annihilated. Most of this energy is lost in the form of energetic neutrinos and gravitational waves that cannot be detected with current instrumentation. The visible light of the supernova outburst comes from the shell of ejected material, which expands outward at 10,000 km/sec or more with a kinetic energy of about 10^{51} ergs.

Some supernova explosions leave behind a compact remnant of 1.4 solar masses with a radius of only about 10 km—a star composed largely of neutrons, whose mean density is 10 times greater than that of a typical atomic nucleus. Conservation of angular momentum requires that a neutron star formed from the core of a star that is rotating, even if slowly, must rotate with a period measured in milliseconds. Compression of any magnetic field embedded in the original stellar core results in a field strength of 10^{12} gauss or more.

Some theorists have proposed that the electromagnetic energy available from the rapidly spinning, highly magnetized neutron star may be responsible for ejecting the outer layers of the star to form a supernova, but in any case, long after the explosion has taken place, intense electric fields are thereby generated, which can accelerate particles to extremely high energies; they in turn emit beams of synchrotron radiation that sweep past the Earth each rotation period to produce the flashes that are characteristic of a pulsar. Hundreds of pulsars have been found with radio telescopes. The pulsar in the Crab nebula, which is the remnant of supernova 1054, has been observed over the entire electromagnetic spectrum from radio waves to gamma rays. The known origin of the nebula in a supernova explosion together with the indication from the observed synchrotron radiation that electrons are accelerated by the central pulsar strongly suggest that supernovae are a prime source of Galactic cosmic rays.

The stellar envelope ejected by the supernova explosion drives a shock wave hundreds of parsecs into the interstellar medium. Such shock waves appear to dominate the dynamics and thermodynamics of the interstellar medium, heating the gas to x-ray temperatures, accelerating interstellar clouds, destroying interstellar dust grains, and in some cases initiating the collapse of molecular clouds to form stars. They also compress the magnetized interstellar plasma and accelerate interstellar ions and electrons to relativistic energies. Radio and x-ray astronomers find evidence for this process in the fact that there is strong synchrotron radio emission from those regions where the x-ray emission shows that the gas must have been heated by a shock wave. Recent theoretical models based on scattering and acceleration of fast particles by turbulence in postshock plasmas can account for both the flux and energy spectrum of Galactic cosmic rays seen at the Earth. The isotopic composition of cosmic rays differs strikingly from that of materials in the solar system, suggesting that at least some cosmic rays originate directly in heavy-element-rich supernova envelopes. Measurements of the relative abundances of radioactive nuclei in cosmic rays, which have a variety of mean lifetimes, have shown that cosmic rays are contained for a few million years in the Galaxy; the constraints this puts on the required energy sources are consistent with acceleration in supernova shock waves.

Binary Star Systems

A large fraction of all stars is found in binary systems. Observations of the interactions between the two members of a binary system permit one to infer properties of the component stars that could not be determined otherwise. For example, precise timing of the radio pulses from a pulsar in a binary system has provided an accurate determination of the mass of the neutron star that causes the pulsar phenomenon; it has also revealed for the first time a slow decrease in orbital period, which theorists predicted to result from the emission of gravitational radiation.

Type I supernovae, whose optical luminosity decreases slowly with time, are probably descendants of white dwarfs in binary star systems of low mass. According to a currently favored model, they occur when the mass accreted from a companion star causes the mass of a white dwarf to exceed the Chandrasekhar limit of 1.4 solar masses, sending it into collapse and causing a thermonuclear explosion. The white dwarf is completely disrupted by such an event, which produces about 1 solar mass of iron-group nuclei, principally the radioactive nucleus ^{56}Ni. The weak decays of ^{56}Ni and of its daughter, ^{56}Co, produce most of the luminosity of Type I supernovae and explain the exponential decline of their light curves. Observation of strong iron lines in the spectrum of supernova 1972e, indicating a large overabundance of iron, lends support to this idea; moreover, a feature attributed to radioactive ^{56}Co has been tentatively identified.

One of the most important developments of the 1970's was the discovery by the Small Astronomical Satellite-1, or *Uhuru*, spacecraft that compact, high-luminosity x-ray sources are located in binary star systems. Their x-ray emission results from the heat generated in the transfer of matter from the atmosphere of a nuclear-burning star to a close companion that is a collapsed star, such as a white dwarf, neutron star, or black hole. The infalling matter is believed to form an accretion disk—a vortex of gas spiraling into the compact object— having temperatures up to tens of millions of degrees.

One such binary x-ray source, Cygnus X-1, has provided the best evidence for the detection of a stellar black hole—a star that is collapsing to a gravitational singularity as predicted by Oppenheimer and Volkoff in 1939. The theory of stellar evolution predicts that collapse to a black hole is the likely fate of the core of a massive star, so that the Galaxy may well contain many millions of black holes. Although an isolated black hole cannot be detected with current techniques, a black hole that is accreting gas from a close binary companion star may be revealed by the x-ray emission of its accretion disk. The analysis of Cygnus X-1 shows that it is a binary system containing a compact object of about 10 solar masses; this object may be a black hole, as no satisfactory alternative interpretation of x-ray emission from the system has been found. Astronomers will continue to search for definitive evidence that Cygnus X-1 and possibly other binary x-ray systems actually do contain black holes.

Many x-ray sources in binary systems were shown to be x-ray pulsars by satellite observatories operating during the 1970's. These systems contain rotating neutron stars with strong magnetic fields that channel the material accreting from a binary companion into narrow columns at the magnetic poles of the star, which are not aligned with the rotation axis. Thermal x rays from the hot polar regions are emitted in broad beams, which sweep around like those of a lighthouse, leading to the observation of x-ray pulses by distant observers. The interpretation of features in the high-energy spectra of some pulsars as electron cyclotron resonances implies that the magnetic fields involved exceed 10^{12} gauss. Precise measurements of changes in the x-ray pulsation rates combined with optical measurements have yielded accurate values for the orbital parameters and the masses of the component stars; they have also provided insight into the dynamics of accretion flows and the properties of ultradense matter in neutron-star interiors.

X-ray bursters form another major class of x-ray binaries. Found in the central regions of our Galaxy and at the centers of condensed globular clusters, these systems emit up to 10^{39} ergs of x rays in bursts lasting only 10 to 100 sec. These bursts appear to originate in weakly magnetic or nonmagnetic neutron stars that are members of close binary systems of low mass. Material accreted by the neutron star accumulates to a critical depth, then undergoes flash thermo-nuclear burning, releasing energy in a burst of thermal x rays; after several hours of renewed accumulation, a fresh layer of material produces the next flash.

A dramatic discovery of the 1970's is the gamma-ray bursts—bright, irregular flashes of gamma rays lasting only a few seconds. Although none of the sources of gamma-ray bursts has been firmly identified, their distribution over the sky suggests that most of them must be relatively nearby. However, the most intense gamma-ray burst ever recorded was observed on March 5, 1979, from the direction of a known supernova remnant in the Large Magellanic Cloud, an irreg-ular galaxy. The intrinsic luminosity of the burst, if it really originated at the 50,000-parsec distance of the Cloud, would be so great as to challenge explanation by any mechanism now known to astro-physics.

Another startling discovery of the 1970's is the remarkable Galactic object SS 433, which apparently ejects more than an Earth-mass every year in two oppositely directed narrow jets of gas moving at ap-proximately one fourth the speed of light. Analysis of periodic changes in the optical emission-line spectrum indicates that the jets precess like a spinning top, tracing out a complete cone every 164 days. This interpretation has been confirmed recently by high-resolution radio images of SS 433 obtained by the VLA and by very-long-baseline interferometry (VLBI), which show the moving helical patterns sprayed into interstellar space by the precessing jets. SS 433 is thought to be a binary system containing a compact object; however, the origin of the system, the mechanisms that accelerate and collimate the jets, and the cause of the precession are still unclear.

Active Galaxies and Quasars

The pioneers of radio astronomy discovered a source of unusual activity in the constellation Sagittarius. That source, Sagittarius A, is now known to be located at the exact center of our Galaxy. VLBI observations have shown that Sgr A is smaller than the solar system, although it is 70 times as luminous as the Crab nebula. Optical studies of the Galactic center region are prevented by interstellar obscuration, but observations of infrared emission, which penetrates the dust, show that it probably contains a massive cluster of red supergiant stars. High-resolution infrared spectra have revealed an emission line of ionized neon at 12.8-μm wavelength arising from clouds of ionized gas in the vicinity of Sgr A. If the Doppler broadening of this line is due to the revolution of the emitting gas about a central gravitating object, the region within about 1 parsec of the Galactic center must contain some 10^6 solar masses of material. While this may be explained by a dense concentration of stars, its association with Sgr A makes it more likely that it is a single massive object, whose accretion disk produces the relativistic electrons responsible for the radio emission of Sgr A. If so, the object may be a massive black hole.

Gamma-ray detectors carried on balloons and on the High-Energy Astronomical Observatory-3 (HEAO-3) spacecraft have observed a strong emission line at 511 keV due to electron–positron annihilation at the position of the Galactic center. Recent HEAO-3 observations show that this line diminished in strength by more than a factor of 2 within a 6-month period, showing that the source of positrons must be extremely compact.

Energetic phenomena similar to those at the center of the Galaxy are observed in other galaxies, but because of their much greater distances they are much more difficult to resolve spatially. For example, there is a spike in surface brightness at the nucleus of the Andromeda Galaxy, Messier 31, which is believed to be caused by a dense concentration of stars. Some spiral galaxies, called Seyfert galaxies, exhibit violent activity in their nuclei, including radio and x-ray emission, rapid motion of ionized gas, and powerful infrared emission concentrated within a region 100 parsecs or less across. The variability of the x-ray emission indicates that it originates in regions less than 1 parsec across. Some elliptical galaxies, called radio galaxies, exhibit similar phenomena, and in addition display enormous radio lobes extending up to 500,000 parsecs into intergalactic space. These lobes appear to be fed by jets of relativistic particles emerging from an active galactic nucleus. For example, Messier 87, a giant elliptical radio galaxy in the Virgo cluster, has a bright, compact optical and x-ray source at its nucleus, from which a jet emerges. Optical studies indicate that there is a concentration of about 10^9 solar masses of material within its central 100 parsecs, a phenomenon that cannot be accounted for by a cluster of normal stars. The nucleus of Centaurus A, another giant radio galaxy, emits a large fraction of its power as gamma rays.

Quasars are starlike objects having large red shifts—the largest value so far observed is 3.5, corresponding to a recessional velocity of 91 percent of the speed of light. The first quasars were discovered through analysis of the spectra of optical counterparts of pointlike radio sources, but it is now clear that only a small fraction of quasars are strong radio sources. Many new quasars have been found recently through spectroscopy of the optical counterparts of faint x-ray sources discovered by the *Einstein* x-ray observatory. While their x-ray, radio, and infrared emission and their optical spectra qualitatively resemble those of Seyfert-galaxy nuclei, quasars radiate much more power.

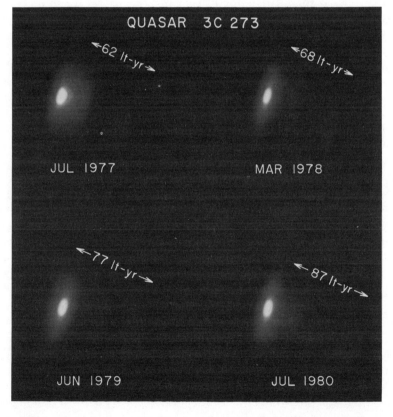

VLBI *radio images of the quasar 3C 273, showing expansion at velocities apparently greater than the speed of light. (Photo courtesy of M. H. Cohen, California Institute of Technology)*

The interpretation of the large red shifts of quasars has been debated for some time; however, the recent demonstration that many quasars lie in the same directions as groups of galaxies having the same red shift strongly supports the conclusion that quasars are really at the vast distances implied by a cosmological interpretation of their red shifts. At these distances, the energy released by quasars is equivalent to the complete conversion into energy of a solar mass of matter every year. Except for the big bang itself, quasars are the most powerful explosions in nature.

Quasars resemble radio galaxies in that compact, active regions are connected by jets to distant radio lobes. The basic energy source of quasars must be located in the compact region (which may well be located in the nucleus of a galaxy); the relativistic particles generated there stream out in jets to the outlying lobes. VLBI observations of the active regions of quasars reveal smaller, jetlike structures aligned with the much larger jets seen with conventional techniques; within these small jets, blobs appear to move outward with speeds that are close to that of light. In some quasars, the transverse velocity of these blobs apparently exceeds that of light; this observation can be explained if the blobs are moving nearly toward us at slightly less than the speed of light. Both the spectra and the time variation of the compact radio sources are consistent with impulsive injection of particles into, and subsequent rapid expansion of, magnetic trapping regions.

X-rays have been detected from many quasars by the *Einstein* x-ray observatory. Many of the known quasars are strong x-ray sources, and new quasars have been discovered solely through their x-ray emission. The x-ray spectra are hard, implying that the x rays are either emitted thermally by hot plasmas or emitted nonthermally by relativistic electrons as synchrotron emission or inverse-Compton scattering. Some x-ray quasars vary on time scales as short as hours, so that the source regions can be no larger than the solar system, despite the fact that the power radiated in some cases exceeds 10^{12} solar luminosities. The number of distant, x-ray emitting quasars is so large that they must account for a substantial fraction of the previously unresolved x-ray background radiation at energies of a few keV. At least one quasar, 3C 273, is known to be a strong gamma-ray source as well as an x-ray source.

The optical spectra of quasars are rich in information, with power-law continua, many narrow absorption lines, and broad emission lines characteristic of rapidly moving clouds of hot gas. In many quasar spectra, the continuum varies with time but the emission lines do not, suggesting that the continuum radiation originates in a much smaller volume. The emission-line red shifts are indicators of the cosmological distances to the quasars. Emission lines include the Lyman lines of hydrogen, which can be radiated by gas of high density, and coronal emission lines, normally radiated only by gas of low density.

As the absorption-line systems usually have lower red shifts than the emission lines, they are attributed to intervening clouds of gas. In one class of absorption-line systems, whose red shifts are only moderately lower than those of the emission lines, heavy-element abundances are normal; such systems probably correspond to clouds of gas ejected by the quasar toward the observer. In the second class of absorption-line systems, the red shifts are considerably lower; as they contain no spectral lines other than those of hydrogen, these systems may arise in intervening intergalactic gas clouds having a primordial chemical composition.

Theoretical models of quasars postulate a layered structure surrounding a compact central energy source. Furthest out are the clouds of cool gas that were ejected by the quasar at earlier epochs and that are now detected through observations of their heavy element-rich absorption lines. Inside is the region where the emission lines are formed. In order to explain the widths of these lines, the source plasma is presumed to be concentrated in clouds that orbit a central region containing a large mass. An intercloud medium, whose temperature is indicated by coronal emission lines to be a million degrees, prevents the clouds from expanding. At the center is a compact energy source responsible for the acceleration of particles and for the generation of gamma-ray, x-ray, optical, and perhaps infrared continua by a combination of thermal and nonthermal processes.

Because of the enormous amounts of energy produced—much of it in the form of relativistic particles—in such a small volume of space, the nature of the central energy source of quasars is of exceptional interest. Among the possibilities are a compact cluster of neutron stars or stellar black holes or both undergoing frequent collisions, a massive plasma cloud stabilized by rotation, and an accretion disk formed from matter spiraling in toward a single massive black hole containing 100 million solar masses or more.

Do quasars and active galaxies really harbor supermassive black holes at their centers? If so, how did they form, and what is the source of matter that feeds the black hole? How is the gravitational energy of the accreting matter converted into the observed radiation? These are among the most exciting unsolved problems in astronomy.

The Impact of Recommended Programs and Facilities on the Study of Violent Events

Current theories of violent events in the cosmos, ranging from solar flares to quasars, present a variety of possible explanations for these events. As explained below, the data collected with the observational facilities recommended in this report will help to choose among the possibilities and thus to advance our understanding of these phenomena.

The basic physical mechanisms that underlie the particle acceleration processes operating in solar flares—mechanisms believed to operate in many other astrophysical contexts—will be investigated with a powerful assembly of instruments aboard the Advanced Solar Observatory (ASO) in space. The Solar Optical Telescope (SOT), which is at the heart of ASO, is planned to be a 1-m-class diffraction-limited telescope that at 5000-Å wavelength can resolve features as small as 0.1 arcsec, or 70 km on the Sun, thus yielding information of unprecedented resolution on magnetic-field configurations before and after solar flares. Extreme-ultraviolet (EUV) and x-ray telescopes of high spatial and spectral resolution on ASO will make measurements of temperatures, densities, and energetic particle populations in the flare plasma. Gamma-ray detectors on ASO will provide information on the numbers and interactions of high-energy particles accelerated by the flare; before ASO is realized, the Gamma Ray Observatory (GRO) will be available to make measurements of the emission of gamma rays by solar flares.

Solar flares radiate strongly in the ultraviolet (UV), the EUV, and the x-ray regions of the spectrum. Flares on other stars can be studied in these regions by ST, the Extreme Ultraviolet Explorer, and AXAF in the forthcoming decade. Elucidation of the nature of stellar flares and their relation to solar flares will require a concerted program of synoptic observations involving these satellites, as well as ground-based optical telescopes.

Important questions regarding the sources, acceleration, and propagation of cosmic rays in the Galaxy will be addressed by new instruments. The precise measurements of elemental and isotopic composition of the cosmic rays as functions of energy, required to give insight into their origin and propagation in the Galaxy, will become possible with the deployment of sensitive detectors on long-duration flights as recommended here. The distribution and spectrum of cosmic rays in the Galaxy can be inferred from the detection by GRO of gamma rays resulting from the interaction of cosmic rays with the interstellar gas and radiation fields. GRO studies of gamma-ray emission from supernova remnants will provide crucial tests of theories for the origin of cosmic rays in supernova explosions and their acceleration by shock waves.

Theories for the formation of heavy elements in supernova explosions will also be tested by GRO; the characteristics specified by the Space Science Board's Committee on Space Astronomy and Astrophysics (CSAA) in its 1979 report would permit it to detect gamma-ray emission lines from freshly synthesized nuclei in supernova explosions out to the Virgo cluster, where a new supernova occurs about once every year. The dynamics of such explosions can be studied by observing the development of their x-ray and UV spectra with AXAF and ST, respectively.

The phenomena displayed by compact stellar objects are so rich and varied that their study will require the concerted use of many of the proposed instruments. The VLB Array will observe the dynamics of jets emerging from Galactic objects such as SS 433 with unprecedented precision. Because of its many independent baselines, it will observe the jets with an order of magnitude greater contrast than heretofore possible. The mechanisms for particle acceleration and radiation in pulsars will become clearer from observations of their x-ray and gamma-ray emissions with AXAF and GRO; for example, the observation of cyclotron lines leads to a direct determination of the magnetic-field strength of the pulsar. AXAF and the X-Ray Timing Explorer (XTE) satellite will advance our understanding of accretion flows onto compact objects, the physics of matter and radiation in superstrong magnetic fields, and the interior structure of neutron stars by making possible studies of fainter objects, by yielding greater precision in determination of spectra, and by making possible millisecond timing studies of variable x-ray sources. Because of their much greater ability to detect faint sources of x rays and light, respectively, than previous systems, AXAF and ST will permit us to observe binary x-ray sources and optically identify them in nearby galaxies, providing clues to their origin and evolution.

The phenomena of transient x-ray and gamma-ray sources will be observed with unprecedented sensitivity by GRO, as well as by XTE and other Explorer satellites. XTE will be able to detect x-ray bursts and transients quickly and to observe their spectral evolution in detail; this is a key clue to the mechanisms of x-ray bursts. An outstanding challenge is to detect and identify the optical and x-ray counterparts of the mysterious gamma-ray burst sources; this will require a concerted program of observations involving GRO, as well as XTE and other x-ray telescopes in space and optical telescopes on the ground.

Many of the new instruments recommended in this report will have powerful capabilities for advancing our knowledge of quasars. Are they really located within galactic nuclei? As explained earlier, ST will answer this question by detecting the parent galaxies of quasars unambiguously. ST images will also permit classification of the quasar-associated galaxy as a spiral, elliptical, or other type, furnishing key information for theoretical quasar models. The increased sensitivity of ST will allow searches for small groups or clusters of galaxies surrounding quasars out to much larger red shifts; an optical telescope with the power of NTT will be required to measure their red shifts.

It is not known why there are no known quasars with red shifts greater than 3.5. Did quasars simply not form until 2 billion or 3 billion years after the big bang, or are more distant ones being obscured by intervening dust? AXAF can detect quasars at red shifts greater than 3.5 in spite of any absorption by dust and determine their red shifts if the iron emission line is present in their x-ray spectra, and thus answer this puzzling question.

Our understanding of the structures responsible for absorption lines in the spectra of quasars will advance greatly as a result of observations of their UV absorption spectra with ST and the far-ultraviolet spectrograph in space. These instruments can detect very low column densities of gas in the outlying regions of intervening galaxies and clusters of galaxies or in intergalactic gas clouds. Such spectra of bright quasars will be good enough to determine physical conditions in the clouds. Knowledge of such conditions should determine whether the absorption-line systems arise from clouds ejected by the quasars or from intergalactic clouds; if the latter, observations in the far ultraviolet will yield the primordial helium abundance, a key datum for cosmology.

ST and the far-ultraviolet spectrograph in space will yield a vast array of new information on the UV emission-line spectra of quasars, so far observed only in the brightest quasars using IUE and sounding rockets. It will be of great interest to find out whether the emission lines in highly red-shifted quasars are the same as those in nearby quasars observed in the optical region. If so, various lines that have been proposed as luminosity indicators can be confirmed and used to determine the distances of quasars at great red shifts, with important applications to cosmology.

The mystery of the central energy source that powers quasars and active galaxies will be addressed in a new way by observations with GRO, SIRTF, and the 10-m submillimeter-wave antenna. For the first time, instruments operating in the gamma-ray region, the far infrared, and the submillimeter spectrum will have sufficient sensitivity to determine the energy emitted in those bands by many quasars and galaxies. This is particularly important because much of the total emission may emerge primarily in those bands. If, as seems likely, the emissions at higher photon energies occur progressively closer to the central energy source, observations of rapid time variations of x-ray emission with AXAF will provide vital clues to the nature of the sources.

Additional clues of a qualitatively different nature will be provided by observations at radio wavelengths. The VLB Array and its extension into space, the first steps of which are recommended in this report, will provide the highest spatial resolution available at any wavelength, about 10^{-9} rad. For objects at a red shift less than 0.1, this corresponds to only 1 light year, and hence there is a chance that one can directly image central radio sources, whose variability indicates sizes of that order. Such observations will provide direct evidence as to how particles are accelerated to high energy and ejected in beams that extend far into intergalactic space.

FORMATION OF STARS AND PLANETS

The Interstellar Medium

Interstellar space is not empty. All along the Milky Way are dark clouds containing microscopic particles of interstellar dust. These clouds are often associated with luminous hot stars, whose lifetimes are so short that they must have formed from the clouds themselves. The interstellar medium is replenished by mass loss from stars as they evolve and die and is thus an essential link in the cycle of stellar death and rebirth that determines the structure and evolution of galaxies.

Between the dark clouds is a relatively thin gas, which in the vicinity of hot stars is ionized and heated (forming H II regions), so that it emits a bright emission-line spectrum. This spectrum shows that the interstellar medium contains hydrogen and helium, along with traces of heavier elements like carbon, nitrogen, and oxygen.

The 21-cm spectral line of neutral atomic hydrogen (H I) has been used to map its distribution throughout our own Galaxy and nearby spiral galaxies. On a Galactic scale, the H I forms spiral arms that coincide with those outlined by dark clouds and young stars. When observers found that individual dark clouds often contain little H I, theorists suggested that the H I in such clouds has reacted on the surfaces of dust grains to form H_2 molecules. Evidence for substantial concentrations of H_2 molecules in dark clouds began to accumulate after 1968 through observations of the microwave spectra of ammonia, carbon monoxide, and other molecules. These data showed that the total density in dark clouds is much greater than that derived from 21-cm observations of H I, strongly implying that molecular hydrogen (H_2) is the dominant material in these clouds.

The *Copernicus* ultraviolet satellite in 1973 confirmed the widespread presence of H_2, at least in regions translucent to starlight, by direct measurements of ultraviolet absorption lines formed by interstellar H_2 in the spectra of distant stars. As predicted, the presence of H_2 is correlated with that of dust. Roughly half of the interstellar hydrogen in our Galaxy (about 5×10^9 solar masses) is H I; the rest is H_2. The far-ultraviolet spectrograph in space recommended in this report will permit extension of the study of H_2 to much greater distances and to darker clouds. Instruments carried aloft by aircraft and balloons will soon make the first observations of the fundamental rotational transition of H_2 at 28.2-μm wavelength. It has been calculated that the enormous sensitivity of a cryogenically cooled infrared telescope in space, such as SIRTF, equipped with array detectors tuned to the 28-μm line, will permit the mapping of interstellar H_2 just as interstellar H I has been mapped in the 21-cm line.

Satellite observations have shown that a broad absorption band at 2200-Å wavelength is present throughout interstellar space. This band is believed to be due to interstellar dust, perhaps carbon in the form of graphite; if so, its strength indicates that a substantial fraction of interstellar carbon is locked up in dust. Complementary observations by *Copernicus* determined the gas-phase abundances of many atoms and ions from their ultraviolet absorption spectra. Atomic carbon is somewhat depleted in the gas phase, consistent with the interpretation of the 2200-Å band; other elements, including silicon, magnesium, and iron, are depleted by large factors, suggesting that they too are locked up in dust grains. That at least some of the interstellar dust is actually composed of these elements has been demonstrated by infrared observations, which reveal strong absorption bands at 10- and 20-μm wavelength, which are characteristic of magnesium-iron silicates. ST and the far-ultraviolet spectrograph in space will extend these studies by covering virtually all of the resonance spectral lines of abundant atoms and ions. The great increase in sensitivity, due to greater aperture, greater multiplexing capability, or both than that of previous instruments, will permit the study of interstellar matter along the lines of sight to more distant and more heavily obscured stars.

With *Copernicus*, astronomers discovered a previously unknown hot (2×10^5 to 5×10^5 K) component of the interstellar medium, in which oxygen is ionized five times (O VI). This medium appears to form a halo around the Galaxy. By reaching stars at great distances, observations of O VI with the far-ultraviolet spectrograph in space will permit us to map the entire gaseous halo of the Galaxy.

Since the late 1960's, radio astronomers have discovered a wide variety of molecules in the interstellar medium. Most of them are concentrated in relatively dense and dark clouds composed primarily of H_2 molecules and dust, called molecular clouds. U.S. radio astronomers have led the world in the study of molecular clouds as a result of the availability of suitable radio antennas and good millimeter-wave receivers; the 11-m millimeter-wave antenna on Kitt Peak operated by the National Radio Astronomy Observatory has played a particularly important role in such studies. The logical next step in this field will be the construction of a 25-Meter Millimeter-Wave Radio Telescope as recommended in an earlier form in the Greenstein report; its larger aperture, more accurate surface, and drier site will permit studies at higher sensitivity, at higher angular resolution, and at much shorter wavelengths (1 mm). Complementing this step, the construction of a 10-m submillimeter-wave radio antenna at an excellent site, as recommended here, will permit the observation of interstellar molecules in selected atmospheric transmission windows down to 340 μm or less with angular resolution approaching 8.5 arcsec.

The CO molecule has been mapped throughout the Galaxy; it is largely confined to spiral arms and is well correlated with interstellar dust. Photographs and CO maps reveal complexes of molecular clouds up to 50 parsecs long. These complexes, of which there are estimated to be about 1000 in the Galaxy, are usually elongated parallel to the Galactic plane and contain individual clouds 10 to 20 parsecs long. Within individual clouds are found regions of high density ("cloud cores") a parsec or less in diameter. It is in such regions that star formation is taking place.

Interstellar dust is thought to originate in the ejecta of red giants and novae, for these objects often display emission bands characteristic of silicate dust. Apparently the dust condenses as the gas expands and cools; when it is subsequently heated by radiation from the nearby star, it emits the infrared radiation that we observe. Infrared surveys have discovered a variety of sources relevant to star formation, including some that have no optical counterparts but that emit microwave continuum radiation and extremely intense spectral lines of OH and H_2O molecules. The energy source for these objects is a luminous hot star hidden from view by a nearby, dense molecular cloud. Its ultraviolet radiation is largely absorbed by a "cocoon" of dust, which re-emits most of the energy in the infrared. Some of the ultraviolet radiation heats and ionizes the accompanying hydrogen, which emits in the radio domain. The OH and H_2O line emissions are produced by natural maser action, the molecular levels being inverted by collisions, infrared transitions, or both. Dust cocoons are invariably found in the densest parts of molecular clouds, supporting the view that the hot stars responsible for their heating were formed recently from the surrounding cloud. With newly developed infrared spatial interferometric techniques, it is now possible to map the formation of dust around many such stars.

Studies of x-ray emission also contribute to an understanding of the interstellar medium. Rocket surveys have mapped the distribution of nearby interstellar gas at million-degree temperatures, leading to a model in which the solar system is now surrounded by an ancient supernova shock wave. The *Einstein* x-ray observatory has mapped x rays from the gas heated by distant supernova explosions. In the future, the increased energy resolution available with AXAF will permit observations of individual emission lines, thereby permitting the detailed quantitative analyses of supernova shock waves and processes occurring behind them, including the destruction of interstellar dust grains by the surrounding, shock-heated gas. The Small Astronomical Satellite-2 (SAS-2) and COS-B satellite have mapped the gamma rays resulting from the decay of pi mesons produced by collisions of cosmic rays with interstellar matter, permitting the distribution of cosmic rays in the Galaxy to be surveyed and the masses of a few interstellar clouds to be determined. The higher sensitivity and angular resolution to be provided by GRO will permit many more of these clouds to be studied at gamma-ray wavelengths.

Our current understanding of the interstellar medium is based on the fact that the energy radiated by interstellar matter must be supplied by the absorption of energy from supernova shock waves, from ultraviolet radiation from hot stars, and from interactions with cosmic rays. Diffuse clouds of atomic hydrogen are maintained at about 100 K by the ejection of energetic electrons from dust grains by ultraviolet stellar photons. Molecular clouds are much colder, on the order of 15 K, because ultraviolet photons cannot penetrate the large amounts of dust present, leaving cosmic-ray interactions as the only source of heat, and because the molecules in such clouds are efficient radiators even at low temperatures. The hot intercloud medium revealed by *Copernicus* observations is maintained at a temperature of several hundred thousand degrees by shock waves from supernovae. The interstellar medium is thus composed of at least three phases having widely differing temperatures. What are the relations between these phases? How do they exchange mass and energy? The capability of the optical, ultraviolet, radio, infrared, and x-ray telescopes of the future to observe all the relevant energy inputs and outputs will greatly increase our understanding of these questions.

Molecular Clouds and Star Formation

The fundamental rotational transition of the carbon monoxide molecule CO at a wavelength of 2.6 mm is a widely used tracer of the distribution of molecular clouds in the Galaxy. An important frontier for future research is to understand the distribution of such clouds in other galaxies; the 25-Meter Millimeter-Wave Radio Telescope (at 2.6-mm and 1.3-mm wavelengths) and the 10-m submillimeter-wave antenna (at the wavelengths of higher CO rotational lines) will permit such studies with spatial resolutions better than 60 parsecs at Messier 31.

A key theoretical problem is the mechanical equilibrium of molecular clouds. Observations of CO and other molecules lead to estimates of the total density and thus the gravitational forces within a cloud. The densities and temperatures derived from CO yield estimates of the gas pressure. In most clouds gravitational forces are much stronger than pressure forces, so that one would expect them to collapse nearly in free fall, a process that should take a few million years. If such collapses lead to star formation, the number of young stars in the Galaxy would be far larger than observed. It is therefore believed that molecular clouds are perhaps prevented from collapsing by internal turbulence, which is known to be present from the observed velocity broadening of molecular lines. Although in many clouds the force exerted by turbulent pressure is adequate to resist gravitational collapse, the turbulence must be highly supersonic and should dissipate rapidly through shock heating. Dissipation can be reduced if molecular clouds are actually composed of large numbers of subclouds, too small to resolve with present instruments. The 25-

Meter Millimeter-Wave Radio Telescope and the submillimeter-wave radio antenna may permit us to resolve such subclouds.

The chemistry of molecular clouds determines their cooling rate and thus their ability to collapse. Chemical reactions among the elements C, N, O, and H under the nonequilibrium conditions that apply in interstellar clouds are of interest in their own right and also because of their possible bearing on the origin of life. Synthesis of molecules in space is thought to begin with the formation of H_2 on the surfaces of dust grains and the subsequent ionization of H_2 by cosmic rays to form H_2^+. Collisions of H_2^+ and H_2 lead to H_3^+, which reacts with C, N, and O to produce species such as HCO^+, CN, HCN, and H_2CO. The complete chain of reactions will be much better understood by observing molecules that have been predicted theoretically, using the increased wavelength coverage of the 25-Meter Millimeter-Wave Radio Telescope and the 10-m submillimeter-wave radio antenna.

Massive stars form in the dense cores of molecular clouds, as indicated by the observation of infrared sources, H_2O and OH masers, and microwave-continuum sources in these regions. According to one theory, dense cores of clouds form when the ultraviolet ra-

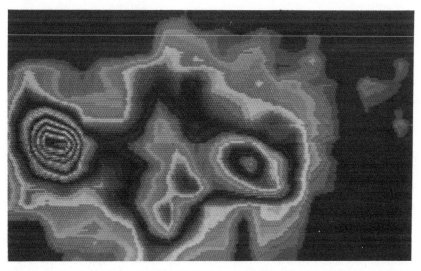

A 12.4-μm map of the central region of the Orion molecular cloud, including the Becklin-Neugebauer "protostar" and the Kleinmann-Low nebula, made with the Goddard Space Flight Center infrared CID array camera. (Photo courtesy of the National Aeronautics and Space Administration, the University of Arizona, and the Harvard-Smithsonian Center for Astrophysics)

diation from massive stars of a previous generation ionize that part of the parent molecular cloud nearest to them, raising the gas pressure and driving a shock wave into the cloud. The shock compression raises the density above the critical value for gravitational instability, and the gas contracts to form a core. Part of the core collapses to form a new generation of massive stars, and the cycle is repeated. In this way, a wave of successive star-formation events propagates through an elongated molecular-cloud complex. This picture is supported by the fact that associations of young stars are observed outside the molecular cloud and aligned with it; the younger associations are nearer the cloud and the older ones further away, as expected from the model. Molecular-cloud cores containing massive young stars are often found immediately adjacent to H II regions. The increased angular resolution available with the 25-Meter Millimeter-Wave Radio Telescope and the 10-m submillimeter-wave radio antenna will help to clarify the spatial relationships involved.

The Solar System

The NASA program of solar-system exploration by deep-space probes is one of the grand technological and scientific adventures of our generation. The opportunities to fly by or to orbit planetary bodies, to study their properties at short range, to make detailed measurements of their atmospheres and magnetospheres, and in some cases actually to land upon their surfaces, have given birth to the new field of planetary science, which draws on a number of scientific disciplines in addition to planetary astronomy. These studies relate strongly to the rest of astronomy and astrophysics. For example, the study of planetary atmospheres has led to the development of concepts and techniques that apply more broadly to the dynamics of stellar atmospheres and the interstellar medium; *in situ* measurements of planetary magnetospheres have provided new insights into the processes of magnetic-field reconnection and particle acceleration; the theory of the internal heating of Jupiter-like planets has consequences for more general studies of the collapse and structure of low-mass protostellar and protoplanetary objects; and studies of comets, meteorites, and other primitive bodies of the solar system have provided insights into the chemical composition of the original solar nebula that may well apply to interstellar clouds generally.

The mechanisms that led to the formation of the Sun and its planets must be at work in the Galaxy today. Much of what we can learn about the formation of our solar system, both from the NASA planetary-exploration program and from ground-based planetary astronomy, may thus be applied to the more general problem of planetary formation. These studies have, during the 1970's, revolutionized our knowledge of the solar system; here we touch on only a few highlights.

Exploration of the Moon has shown that it is now inactive; its surface is the result of meteoritic cratering over its 4.5-billion-year lifetime. Mercury is also inactive, but Venus, like the Earth, shows the effects of tectonic activity—great uplifts and a very dense atmosphere, presumably of secondary origin. The *Viking* mission showed that Mars is at present inactive and that only small amounts of water are now present on the surface. On the other hand, images from the *Viking* orbiter demonstrate that Mars was once active and may have been washed by water at one time.

It has recently become possible to classify asteroids by their infrared reflection spectra. The evidence suggests that an important class of meteorites, the carbonaceous chondrites, must be derived from certain classes of asteroids, presumably through collisions that injected them into Earth-crossing orbits. Because of the large collecting area of NTT, it will extend studies of infrared reflection spectra to much smaller objects.

The *Voyager* program has revolutionized our knowledge of Jupiter, Saturn, and their magnetospheres and satellites. With tremendous detail, it has revealed complex motions in Jupiter's deep atmosphere, verified its hydrogen-rich composition, and discovered enormous atmospheric lightning flashes. The Galilean satellites were found to be complex and completely different from one another; Io is heated to the melting point by Jupiter's tidal force and emits sulfur ions into the Jovian magnetosphere from a number of active volcanos. Future infrared telescopes on Earth and in Earth orbit will contribute to further understanding of all the planets by imaging and spectrally analyzing their infrared emission. NTT will speed up spectroscopy in the near infrared, while the LDR, because of its large collecting area and high angular resolution at 20-μm wavelength and longer (better than 1 arcsec) will permit high-resolution infrared spectroscopy of planets with many resolution elements over the disk. ST will have a unique capability to image Jupiter and its satellites regularly from Earth orbit with the same 150-km spatial resolution achieved only briefly by *Voyager*; in addition, ST will spectroscopically analyze planetary atmospheres using the reflected ultraviolet light from the Sun and the light emitted from upper atmospheric layers. We will gain a better knowledge of the complex chemistry occurring in comets by studying their ultraviolet reflection spectra with ST, and their molecular spectra with the 25-m Millimeter-Wave Radio Telescope and the 10-meter submillimeter-wave radio antenna.

Current theories of the origin of the solar system postulate a "solar nebula," a disk of gas and dust that formed together with the Sun nearly 5 billion years ago. It is believed that dust particles (either interstellar particles brought in when the solar nebula formed or particles that condensed later out of gases in a heated nebula) migrated to its midplane, where gravitational attraction drew them together to form the planetesimals, which, through collisions, accumulated to form the terrestrial planets and the cores of the outer planets. This model has been severely constrained by the recent discovery of isotopic anomalies in meteorites. Not only must the whole process have occurred faster then previously thought possible, but ^{26}Al must have been injected just before the process started. Since both the supernova explosion responsible for producing the ^{26}Al and the origin of the solar system were rare events, it seems likely that they were causally related, perhaps because a nearby supernova explosion triggered the collapse of a molecular cloud. Observers have recently reported configurations of young stars that may have formed as the result of compression by such a supernova shock wave.

Roles of Theory and Observation

The formation of stars and planets is a formidable theoretical problem. What mechanism stabilizes molecular clouds? Are supernova shock waves required to make them unstable toward collapse? Why do they fragment into objects of stellar mass as they collapse, and what determines the distribution of masses? How does the angular momentum observed in the solar system and in binary-star systems originate? What role do magnetic fields play in the collapse and fragmentation process? Why do some fragments become planetary systems, while others become binary-star systems or isolated stars? What determines the distribution of planetary masses and distances? Do satellites form at the same time as the planets, or later?

Such questions are now receiving serious theoretical study. Full two- and three-dimensional hydrodynamic computer codes, taking proper account of the effects of turbulence and magnetic fields, are needed for satisfactory answers. Such codes, originally developed for other applications, can now be implemented only on supercomputers, but the combination of array processors with the smaller computers recommended in this report may also be able to handle hydrodynamic codes for the formation of stars and planets.

Future observational studies should aim to constrain theoretical models at every stage. The question of turbulent stabilization of molecular clouds requires the greatest possible angular resolution in order to determine whether the observed velocity broadening is actually due to the overlapping velocity shifts of various subclouds. Both the 25-Meter Millimeter-Wave Radio Telescope and the 10-m submillimeter-wave radio antenna offer dramatic improvements in this area, providing spatial resolutions down to 0.02 parsec in the Orion molecular cloud.

In the study of dense cores of molecular clouds, it is important to obtain data in the spectral lines of a number of different molecular species that depend differently on temperature and pressure and, because of differing saturation effects, arise in different parts of the core. The 25-Meter Millimeter-Wave Radio Telescope and the 10-m submillimeter-wave radio antenna will be able to detect many more lines because of the extended frequency range. It is also important to extend the search for young stars in these cores to much fainter limits than is now possible, in order to assess whether fragmentation results in simultaneous formation of a range of stellar masses. SIRTF, with its 1000-fold increase in infrared sensitivity, will be crucial in detecting much fainter stars.

Because of its relation to complex molecules, the transformations of carbon are of particular interest; in diffuse clouds, carbon may be divided among neutral atoms, ions, and dust particles. The 10-m submillimeter-wave radio antenna will permit the mapping of neutral atomic carbon, the only atom other than H I for which this is possible. In molecular clouds, extension of the accessible frequency range using the 25-Meter Millimeter-Wave Radio Telescope and the 10-m submillimeter-wave radio antenna will bring many new organic molecules, as well as excited states of known molecules such as CO and HCN, under study. In the vicinity of young stars and of shock waves formed in cloud cores, the temperature is high enough to excite submillimeter and infrared lines of molecules. The 10-m submillimeter-wave radio antenna will permit spectroscopic observations of many such lines at wavelengths down to 340-μm wavelength, while optical-infrared telescopes, including NTT, will work up to 20-μm wavelength. The far-infrared range between these wavelengths can be studied with moderate-sized instruments on the Kuiper Airborne Observatory and on balloons, and the LDR will be able to do high-spectral-resolution work at arcsecond resolution on these objects.

A protoplanetary disk like that of the solar nebula, located at the 170-parsec distance of the Ophiuchus dark cloud, would subtend 0.3 arcsec if it is 25 astronomical units in radius. If its temperature is 100 K, it would radiate in the 20–50-μm wavelength range, with a total luminosity about equal to the Sun. Such an object is beyond current ability to detect but would be easily detected by SIRTF. A LDR in the 10-m class would be able to study the spectra of such objects because of its large collecting area; it would achieve some spatial resolution on the object when operated with low spectral resolution.

If current ideas are correct, planetary systems are ubiquitous. So far, however, attempts to confirm the astrometric detection of planetary-sized objects orbiting Barnard's star have failed, leaving us with no conclusive evidence for the existence of any planetary system but our own. This problem is of such importance that new efforts should be made to solve it, as explained later in this chapter in the section on Planets, Life, and Intelligence.

SOLAR AND STELLAR ACTIVITY

Activity on the Sun

It has been known for a century that there are gigantic eruptions on the Sun, with conspicuous terrestrial consequences like the Aurora Borealis and radio blackouts. The solar atmosphere is the site of intensely energetic and extraordinarily complex activity, much of which is still not thoroughly understood. Insight into these phenomena is important for our understanding not only of the Sun but also of other stars, as well as of plasma processes in a wide variety of astronomical objects.

The magnetic field of sunspots was discovered 70 years ago. The solar corona has long been known from observations during total eclipses of the Sun, but not until 40 years ago was it demonstrated that its temperature is a million degrees. Gradually, as instruments improved, additional features of solar activity were discovered, including eruptive prominences, small flares that appear in newly emerging active regions, and occasional large flares. It was found that flares are the source of highly energetic particles, called solar cosmic rays. The general magnetic field of the Sun was discovered, mapped, and followed throughout the entire 11-year sunspot cycle using the newly invented magnetograph. The intimate connection of magnetic fields with solar activity—super-heated gases, fast particles, violent eruptions, and flares—became clear.

The motion of material in comet tails provided the first hint that the solar corona is expanding. Experiments on Earth-orbiting spacecraft revealed that the expansion of the corona causes a supersonic solar wind that carries away a million tons of solar material every second at velocities between 300 and 600 km/sec. In the 1970's, instruments aboard Skylab discovered that the corona erupts every few hours, and x-ray studies from that satellite outlined in dramatic detail the hot, dense regions of the active corona and their association with the solar magnetic field, as well as the cooler, more rarefied regions called coronal holes, from which high-speed solar-wind streams flow. Skylab also discovered x-ray bright points, from which most of the magnetic flux of the Sun seems to emanate.

The solar wind extends far into space, forming a "heliosphere" within the surrounding interstellar medium, which is a vast sea of plasma activity perhaps several hundred astronomical units across. It is filled with fast and slow streams of solar wind, interacting violently to produce shocks and fast particles, and is swept constantly by blast waves from solar flares. Interaction of this plasma with planetary magnetospheres produces planetary magnetic storms and radiation belts and accelerates particles to high energy near the planets. The subject of intensive study through direct observations and theoretical modeling, the heliosphere also serves as a guide to the complex phenomena that must also occur near other stars, although they are too distant for their tenuous plasma envelopes to be observed.

In recent years, historical studies have revealed that the activity of the Sun may fall to a low level for a century at a time, as it did during the fifteenth and seventeenth centuries, or rise to a very high level, as it did in the twelfth century. The climate of the Earth appears to respond to these long-term variations, with the mean annual temperature in the Earth's North Temperate Zone declining in extended periods of low activity. The causes of such long-term solar variations remain a mystery.

The scope of solar activity is impressive; nothing like it is known in the terrestrial laboratory. How are magnetic fields produced inside the Sun? How do these fields cause flares, eruptive prominences, solar cosmic rays, and sunspots? How do they control and perhaps heat the active corona? And why does the corona expand to produce the solar wind? Considerable progress has been made over the past three decades in answering some of these questions, while others have evaded understanding to this day.

Detailed observations of the magnetic fields at the surface of the Sun with the precision vacuum tower telescopes and high-resolution magnetographs and spectrographs currently available have demonstrated that the solar magnetic field is concentrated into individual flux tubes in which the field is intense; incredibly, these tubes do not expand despite the huge internal magnetic pressure. The formation of sunspots, plages, and flares and the heating of the corona are profoundly affected by this seemingly unnatural behavior. Renewed efforts with infrared, optical, ultraviolet, x-ray, and gamma-ray telescopes, together with determined theoretical studies, will be necessary for progress toward understanding some of the more complex aspects of solar activity during the 1980's.

Instruments aboard the Solar Maximum Mission carried out the first precise, coordinated studies of flares in ultraviolet light and x rays, observing particularly the extremely compact and intense centers of activity with temperatures above one hundred million degrees that lie at the heart of solar flares. The Solar Optical Telescope (SOT), a diffraction-limited meter-class telescope to be launched on the Space Shuttle during the 1980's, will resolve detail down to 0.1 arcsec, or 70 km on the Sun. For the first time, we will be able to see the structure of the most slender magnetic tubes and the details of the convection around them, as well as the glowing hydrogen filaments above. The SOT will be followed by the Advanced Solar Observatory (ASO) in space, which will provide better than 0.1-arcsec resolution images simultaneously at optical, ultraviolet, and soft x-ray wavelengths. These observations should help to reveal the basic physics of a great variety of phenomena on the Sun, including plages, sunspots and their penumbrae, the breakup of magnetic fields into distinct flux tubes, and related activity at the boundaries of convective cells in the solar atmosphere, because many of these phenomena are expected to show important details as one exceeds the present ground-based resolution limit of about 1 arcsec. Because of the tremendous temperature range from the photosphere (6000 K) to the corona (2×10^6 K), simultaneous observations in all three spectral ranges are needed to establish the connections between phenomena seen in different atmospheric layers. Such insights will provide tests of present and future theoretical ideas that are in many cases much more stringent than can be furnished by the best observations from the ground.

The ASO recommended in this report will be designed to obtain the most detailed observations of the solar surface that can be made from Earth orbit with a mission in the moderate-cost class. Spacecraft launched as part of the International Solar Polar Mission (ISPM) will add an important new dimension to our knowledge of the Sun and of the heliosphere by making observations out of the ecliptic plane. The desirability of obtaining extremely detailed observations of the solar surface raises an exciting possibility—a probe to the Sun itself. NASA's proposed Star Probe mission, planned to carry out the first *in situ* exploration of any star, would come within 2 million kilometers of the solar surface; if it were to carry a 10-cm telescope, it would be able to resolve details of solar activity down to 10 km. The launch of such a mission could open up a new era of investigation in solar physics, much as NASA's program of planetary exploration by deep-space probes has broadened and supplemented the traditional field of planetary astronomy.

Studies of the solar core will receive renewed impetus during the 1980's. It was to probe the deep interior of the Sun that neutrino astronomy was born a decade ago with the assembly underground of a ^{37}Cl neutrino detector in the Homestake Mine in South Dakota. The flux of solar neutrinos detected by this experiment is less than one third as intense as predicted by current theories of stellar energy

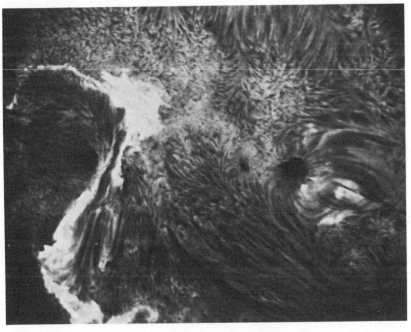

A large solar flare photographed in hydrogen light. The combed appearance of the solar gases is due to the strong magnetic fields in the flare region. (Photo courtesy of the Association of Universities for Research in Astronomy, Inc., Sacramento Peak Observatory)

generation and internal structure. This result has already led to the revision of models of the solar interior; in view of the magnitude of the discrepancy, however, a detector incorporating large quantities of gallium, which is sensitive to the full range of neutrino energies, will be needed to check our understanding of energy production in the Sun. More accurate laboratory measurements of the relevant nuclear reaction rates are also needed to refine the value of the solar-neutrino flux predicted by theory.

The study of long-period oscillations of the surface of the Sun, which arise from waves trapped beneath the surface, provides insight into the properties of the solar layers between the atmosphere and the core. It was recognized during the 1970's that the solar 5-min oscillations are global phenomena that can be used to probe the structure and dynamical behavior of the solar convection zone hidden beneath the surface. Recent observations of these oscillations from a station located in long-term sunlight near the South Pole of the Earth have led to precise determinations of the oscillation frequencies. However, the study of solar dynamics over much longer time periods than the polar observations can provide is essential to the successful completion of such studies. This goal will require measurements from an Earth-orbiting spacecraft, such as a proposed Solar Dynamics Explorer satellite.

Stellar Activity

At about the same time that the magnetic field of the Sun was being mapped by magnetographs, similar instruments showed that many other stars also have magnetic fields; they range up to a hundred or a thousand times more intense than that of the Sun. Strong fields covering much of the surface have been detected directly only in stars more massive than the Sun, but recent evidence implies that less massive stars also have magnetic fields. Observations of chromospheric lines in less massive stars indicate that they, like the Sun, are surrounded by superheated gases, suggesting that they also have active regions and therefore localized magnetic fields. Many of these stars display magnetic cycles, during which stellar activity waxes and wanes with a regular period like the sunspot cycle.

Ultraviolet and x-ray astronomy from space vehicles led to a breakthrough in stellar-activity studies. Observations from the *Copernicus* ultraviolet observatory showed that some stars possess stellar winds; in some cases, the wind is so strong that it must have profound effects on the evolution of the star. In addition, stellar winds carry away so much angular momentum that the rate at which the stars rotate must be slowed, over hundreds of millions of years, to low values. The International Ultraviolet Explorer (IUE) satellite extended these studies to a much greater variety of stellar types.

Observations with the *Einstein* x-ray observatory revealed that stars of nearly all types are surrounded by coronas at temperatures reaching millions of degrees, implying the existence of magnetic fields and hence active regions, stellar winds, and perhaps flares. Some stars exhibit such large "starspots" that one entire face of the star is dimmed; others exhibit flare activity thousands of times more intense than that on the Sun.

The Sun serves as a local laboratory where detailed work can be done to understand the individual physical effects that contribute to

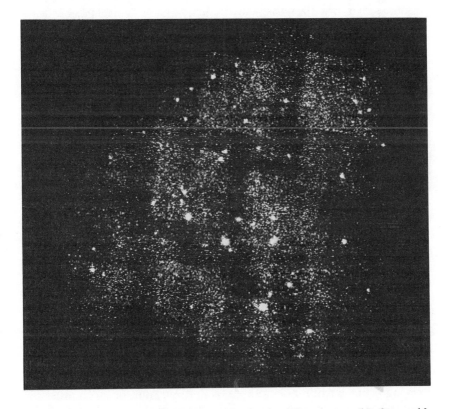

Einstein x-ray image of the young star cluster The Hyades. (Photo courtesy of R. Stern and J. Underwood, Jet Propulsion Laboratory, and M. Zolcinski and S. Antiochos, Stanford University)

this richly varied activity. The exotic and unanticipated properties of magnetic fields in convective fluids are legion; we have only begun to discover what they are and how they fit together to produce solar activity. The Sun offers a close look at the complex detail of magnetic activity, whereas the surfaces of distant stars cannot be studied in detail because stars appear only as points of light in the largest telescopes. On the other hand, recent observations show that activity can be identified and analyzed in other stars using the radiation from the whole star.

While a coordinated attack on problems of solar activity is being mounted, observations of activity in other stars should also be pressed forward with ground-based telescopes, as well as ultraviolet and x-ray telescopes in Earth orbit, including ST, AXAF, and the far-ultraviolet spectrograph in space. Each of these instruments will have far greater sensitivity than its predecessors and, therefore, will be able to study far more stars than have been possible with earlier instruments. The goal is to reach stars with many different characteristics; theoretical predictions of the dependence of activity on the age and rotation period of a star can be tested much better with studies of many stars than they can if only the Sun is studied.

The Role of Magnetic Fields

Magnetic fields lie at the heart of solar and stellar activity. They are somehow produced within most stars, as well as within several of the planets of the solar system. In the Sun, the field floats up to the surface, where it produces sunspots. In dissipating, the field heats the plages and the corona; if it does so rapidly enough, it causes explosions or eruptions, sometimes accompanied by superheated flares.

The origin of magnetic fields in the interior of the Sun and other stars poses a fundamental problem to physics. It has been suggested that magnetic fields were trapped in the interiors of stars at the time they formed from interstellar gas. According to this suggestion, the field slowly rises to the surface of the star over its lifetime. However, the regular 11-year reversal of magnetic field in the Sun, as well as the cyclic variations observed in other stars, appears to require another explanation.

Any primordial magnetic field in the Earth must have dissipated long ago; the Earth's magnetic field that we now observe must be continually regenerated by the flow of electrically conducting fluid in the core. Just as the convection and nonuniform rotation of the liquid iron core is thus responsible for the maintenance of the field of the Earth, convection and nonuniform rotation in the solar convective zone can maintain the field of the Sun.

The generation of magnetic fields is described by the equations of magnetohydrodynamics. Mathematical solutions of these equations incorporating reasonable estimates for the fluid velocities within the convective zone predict magnetic fields remarkably like the actual magnetic fields of the Sun. On the assumption that the angular velocity of solar gases increases with depth, the equations predict that an east–west magnetic-field component develops first at middle and high solar latitudes and then migrates toward the equator, where the oppositely directed fields from opposite hemispheres meet and cancel each other. The north–south fields are predicted to reverse direction roughly when the east–west field reaches tropical latitudes and maximum strength, as is observed.

Although this theory is evidently on the right track, it is based on an untested model of solar convective motions. Little is known about the actual motions of fluid beneath the solar surface or about the complicated hydrodynamics of the internal convection, circulation, and nonuniform rotation of the Sun. Observations of the north–south circulation at the surface, together with determinations of the angular velocity at various depths through observations of the 5-min surface oscillations, are essential for a better understanding of solar convection and magnetic-field maintenance.

The generation of magnetic fields in other stars will remain a puzzle until we have a solid understanding of the circulation in the Sun. The stronger fields in younger stars, implied by their higher levels of activity, may be a direct consequence of their higher rates of rotation. However, it is not at all clear how the very strong magnetic fields of stars more massive than the Sun are produced. Not only are such stars slow rotators, but it is also believed that they have no internal convection. The origin of stellar magnetic fields thus remains obscure.

The often explosive dissipation of magnetic fields and the accompanying acceleration of particles to high energies is a major problem in astrophysics. Laboratory and theoretical work, as well as observations of magnetospheric phenomena in the Sun and planets, shows that reconnection of magnetic-field lines to form configurations of lower energy is a key mechanism. Reconnection requires diffusion of magnetic fields through the gases in which they are embedded; however, the rate of diffusion is negligible according to the equations of magnetohydrodynamics. Apparently, it is necessary to invoke plasma instabilities not included in those equations; once established, they may also facilitate the conversion of magnetic energy into the energy of fast particles. Understanding magnetic reconnection and particle acceleration is essential not only for magnetospheric physics and solar physics but also for other violent events in astronomy, including those associated with x-ray sources and quasars. Only in the Sun and planets can these processes be observed both on the scale and in the detail required for a clear understanding of many of their features.

Stellar Mass Loss

The importance and ubiquity of strong stellar winds became apparent during the 1970's through advances in space ultraviolet astronomy and ground-based infrared astronomy. It is now clear that both the hot, blue giant stars and the cool, red giant stars have stellar winds and that these winds are sometimes prominent features of regions of star formation. The most luminous stars appear to lose mass at rates up to a billion times the mass lost in the solar wind. A luminous giant star may thus lose a substantial fraction of its mass even during its relatively short lifetime of a few million years, with profound effects not only for its subsequent evolution but also for its interstellar environment.

The luminous, hot blue giant stars have by far the strongest stellar winds. Observations of their ultraviolet spectra with telescopes on rockets and on the *Copernicus* and IUE satellites have shown that these winds flow at speeds up to 3000 km/sec and are characterized by temperatures below 10^5 K. The mass-loss rates inferred from infrared and radio observations approach and sometimes exceed 10^{-5} solar masses per year. Such winds cannot be driven by gas pressure alone, as is the solar wind, but must be driven instead by the stellar ultraviolet radiation pressure, scattered from ions in the wind; the same mechanism is thought to play a role in the ejection of gas from galactic nuclei and quasars. These winds are so powerful that they hollow out enormous cavities in the interstellar gas, pushing outward expanding shells of interstellar material that resemble those generated by supernova explosions. The mechanism for driving the winds in hot stars is not well understood. Time variations in ultraviolet spectra of the stars suggest that the wind is unstable, and theorists are investigating possible instabilities in an effort to nail down the characteristics of the underlying flow.

Observations with optical, infrared, and radio telescopes show that cool, red-giant stars have winds comparable in strength with those of the hot, blue-giant stars but with temperatures less than 3000 K and with much lower velocities, about 30 km/sec. No ultraviolet or x-ray emission is seen. These winds are rich in dust grains and molecules; since nearly all stars more massive than the Sun eventually evolve into red giants, such winds provide a major source of new interstellar gas and dust, furnishing a vital link in the cycle of star formation and galactic evolution. As for hot stars, the mechanism for driving these winds is not well understood. Radiation pressure appears to be inadequate to sustain such a high-mass-loss rate, and the gas pressure is too low. Perhaps instabilities, turbulence, and/or magnetic fields in the stellar atmosphere are responsible.

Recent observations of H_2 and CO molecules in the Orion nebula with infrared and radio spectrographs have revealed clouds of gas expanding outward at velocities approaching 100 km/sec. VLBI radio observations disclose expanding knots of H_2O maser emission associated with the region of star formation in the Orion nebula, suggesting that strong stellar winds are associated with protostars. The cause of these winds remains unknown.

The new instruments for the 1980's recommended in this report will have powerful capabilities for investigations of strong stellar winds and their consequences. AXAF will carry out sensitive surveys and spectroscopic observations of x-ray emission from the winds of luminous hot stars, permitting studies of the properties of the coronal gas and the mechanisms for heating it. ST and the far-ultraviolet spectrograph in space will greatly increase the number of stars that are observable over the number observable with *Copernicus* and IUE and furnish a better understanding of the physical conditions in the winds of hot stars, of their driving mechanisms, and of their interaction with their interstellar environments.

There is also much to be learned about the winds of both hot and cool giant stars and the winds associated with regions of star formation from infrared observations with new instruments. Because its large aperture will make possible higher spectral-resolution measurements, NTT will permit us to determine mass-loss rates and velocity profiles of the winds of hot and cool giant stars. Because of its high sensitivity, SIRTF will be able to observe stellar winds and their associated dynamics in much more optically obscured regions of star formation, and the LDR, because of its large aperture, can measure their spectra. The VLB Array will provide more reliable observations of the motions of the radio maser sources in such regions, because its enhanced sensitivity will allow the study of fainter, more numerous sources.

PLANETS, LIFE, AND INTELLIGENCE

Molecular biology has given penetrating insight into the nature of life. Complex DNA molecules in the nucleus of every cell encode all the information required to determine how the cell develops and functions. All the rich diversity of life on Earth is coded in strands of DNA, which have evolved from primitive forms that apparently arose through a series of transformations in our corner of the Universe. As we consider the myriads of stars in the Universe, we wonder what other genetic systems may exist.

Are we alone in the Universe? Questions concerning the origin and fate of life on Earth have been pondered since ancient times. Astronomy has shown that there are enormous numbers of stars like the Sun and that the abundances of chemical elements are much the same everywhere. It seems possible, therefore, that there are habitats for life scattered throughout the Universe. Life on Earth evolved from primitive forms by mutation and natural selection into new species of ever-increasing complexity. The assumption that such processes operate wherever the conditions are right remains speculative until other examples of life in the Universe are found.

Life in the Solar System

Life on Earth is so intertwined with the chemistry of the atmosphere and oceans as to form a single ecosystem, in which each part is affected by the others. Based on carbon, an abundant element, and dependent on liquid water, the ecosystem is sustained by the low-entropy energy of sunlight, captured through photosynthesis. Beyond these basic considerations, it is not clear what other properties of planet Earth played an essential role in the origin of life. Solid surfaces? Gravity? The daily cycle of light and dark?

Several planets in the solar system are similar enough to the Earth that a search for life on them can shed some light on these questions. Mars has long appealed to scientists and the public alike as an intriguing target for exploration. The 1970's have seen a remarkable effort in that direction, culminating in the orbiting spacecraft and landers of the Viking mission. Although the best-publicized results were the photographs from the landing sites and the chemical searches there for microorganisms and organic molecules, large amounts of other data were also returned that relate to the Martian atmosphere and temperature and the geological processes that have shaped the Martian surface. The conclusion that emerges from these studies is that the Martian atmosphere, now dry and cold, was perhaps quite different in the past; probably large amounts of water precipitated at one time, shaping the surface in torrential floods. Conditions at that time could have been more favorable to life, so that further studies are required to tell us whether life ever existed on Mars.

Venus has been the subject of intensive study by both the United States and the Soviet Union. Its surface is hot enough to melt lead; this and the fact that the atmosphere is topped by sulfuric acid clouds

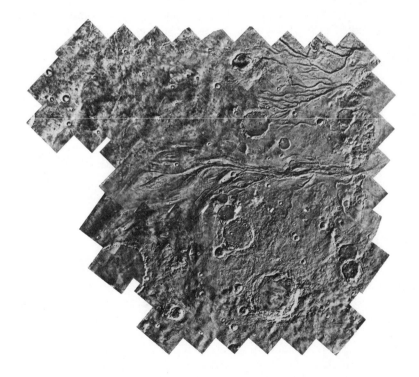

Old stream beds on Mars, west of the Viking landing site, as seen from the Viking Orbiter. The terrain slopes upward to the left. (Photo courtesy of the National Aeronautics and Space Administration)

indicate that Venus is inhospitable to life. Mercury is also too hot during its long days, and it has no significant atmosphere. Beyond Mars, Jupiter has been the subject of speculation because its dense atmosphere contains organic molecules that could constitute the first steps toward the origin of life. The temperature is extremely low at the visible cloud tops, but it increases inward, so that organisms would be quite comfortable at an appropriate depth. However, there is no evidence of a solid or liquid surface for life to cling to—if, indeed, that is essential. The remarkable properties of Jupiter's satellites were revealed for the first time by the *Voyager* photographs. Most appear to be too cold and to lack sufficient atmospheres. Although Io is internally heated, it is overrun by hot lava and subjected to sulfurous volcanic fumes, so that its surface would be hostile to most forms of terrestrial life. Beyond Jupiter, most bodies are too cold for life to be credible, although Titan, the largest satellite of Saturn, is now known from *Voyager* observations to have an atmosphere composed primarily of nitrogen, with a temperature at the surface near the triple point of methane; it is conceivable that interesting organic molecules, of types believed to be precursors to primitive life, could exist there.

In all likelihood, however, life in the solar system has been confined to the Earth and possibly Mars. On Earth, the development of intelligence has led to the explosive development of technology and thus to radio communication and powerful radars; the earliest radio signals have by now propagated 50 light-years into space, past several thousand stars more or less like the Sun. In principle, then, intelligent beings far from our solar system could learn of our existence by studying radio waves from space, just as we could discover any such beings by using similar means.

Conditions for Life in the Universe

The discovery of complex interstellar molecules shows that chemical processes possibly relevant to the origin of life are ubiquitous. Among them are formaldehyde and hydrogen cyanide, both of which play an important role in laboratory experiments directed at recreating the chemical evolution that preceded the origin of life. More than 50 different molecules have now been found in molecular clouds, including long-chain molecules such as HC_9N. The abundances of the simpler molecules observed have been explained by ion–molecule exchange reactions, but it is not yet certain whether the more complex molecules require other processes, such as surface catalysis on dust grains, for their explanation.

When a molecular cloud collapses to form stars and planets, its complex molecules may be destroyed. On the other hand, molecules might survive the extreme conditions encountered during the collapse by condensing onto dust grains. The complex organic molecules found in carbonaceous-chondrite meteorites may have formed in this way. Such a process could be critically important for the origin of life in the solar system, since carbon would otherwise be locked up as either carbon monoxide or methane gas, which are too light to be gravitationally bound by planets as small as the Earth. By condensing on grains, carbon can be bound into planets, where it is available for life.

The synthesis of organic molecules in space will be studied more intensively in the coming decade. By enabling studies of new molecules and higher energy states, with increased sensitivity and angular resolution, the millimeter, submillimeter, and infrared instruments recommended in this report offer an opportunity to pursue the chemistry of carbon in interstellar clouds, and particularly in the collapsing cores of dark clouds, where stars and planets are believed to form. Furthermore, they will offer additional opportunities to study the molecular composition of comets, which are believed to be samples of the most primitive material in the solar system.

Planets like our own may be the most likely habitats for life outside the solar system. However, in view of the lack of confirmation of a planet orbiting Barnard's star, there is as yet no certain evidence that any planets exist beyond our solar system. Current theories of star formation suggest that a fragment of a dark interstellar cloud destined to become a star would have substantial angular momentum, which would set into rapid rotation any gas and dust that accompanies the cloud-to-be-star on its inward gravitational collapse. The accompanying centrifugal force would resist the inward pull of gravity and result in a rotating disk of gas and dust that would provide a natural breeding place for planets, arrayed as ours are, in a great flattened and rotating system.

So much is speculation; to prove that planets exist near other stars, we must observe them. This is a formidable task, given the fact that even the largest planet in the solar system, Jupiter, has only a tenth of the size and a thousandth of the mass of its parent star, the Sun. One way a distant observer could detect Jupiter is through its gravitational pull on the Sun. This effect would cause a variation in the radial velocity of the Sun of a few tens of meters per second and, as observed from a distance of 5 parsecs, a 1-milliarcsecond displacement on the solar position. Could we detect a Jupiter-sized planet around a nearby star? There already exist several techniques for obtaining radial-velocity measurements of the required precision; sustained programs of observations of many candidate stars are now required. Astronomers have expended great effort to make milliarcsecond-position measurements on nearby stars, but until recently the required precision has not been available. With the development of the optical astrometric techniques recommended in this report, however, it should be possible to observe many stars with a precision exceeding 1 milliarcsecond and by this means to detect Jupiter-sized planets around nearby stars, if they exist. Space astrometry should ultimately yield much higher positional accuracy, leading to the still more interesting prospect of detecting Earth-sized planets. Far-infrared interferometric observations from space could also reveal planets around nearby stars.

Search for Extraterrestrial Intelligence

Even if other planets are detected, it will still be difficult to infer whether life is present; to do so directly would require imaging the planet itself with as yet undreamed-of resolution. Only if the planet harbors intelligent life capable of producing electromagnetic signals detectable at Earth is there at present any hope of finding life outside our solar system. It is a remarkable fact that radio and television signals generated copiously on Earth could be detected at distances of many parsecs by civilizations that, like ours, would otherwise have no way of knowing of the existence of details of the planet that is our home.

Should the human race search seriously for signals from other possible civilizations? Much has been written about this question, both on a technical and a philosophical level. Reception of intelligent signals from space could have a dramatic effect on human affairs, as did contact between the native peoples of the New World and the technologically more advanced peoples of Europe. The effects would be beneficial, if the information could be deciphered and should prove generally useful; on the other hand, they could be harmful if humanity is not ready to use the information wisely.

The technology is now available to make significant searches of this kind. The 300-m radio telescope of the National Astronomy and Ionosphere Center at Arecibo, Puerto Rico, is capable of receiving a message beamed at us from any of the hundreds of billions of stars in our Galaxy, provided the civilization sending the message were transmitting with a facility similar to that at Arecibo. Several searches for such extraterrestrial signals have already been undertaken, so far with negative results, but the rate of improvement of communications technology is so rapid that each search has been far more sensitive than its predecessors.

We are entering an era when it is technically possible both to detect planets around nearby stars and to detect signals from intelligent life on planets immensely farther away, even if we cannot detect the more distant planets themselves. Both investigations would bear directly on important scientific questions. Our interest in the tiny fraction of the matter in the solar system that condensed into planets is heightened by the fact that life has developed on at least one of them. Have condensations to planets and the origin of life occurred elsewhere as well? And has that life evolved into communicative intelligence, with which we human beings might be able to enter a conversation about life in the Universe?

These questions reach far beyond astronomy, and even beyond science as we currently think of it. Yet astronomers, who are in a sense commissioned by the public to keep an eye on the Universe, feel bound to ask them and to point out how we might begin to try to answer them. It is for these reasons that the Committee recommends that in the 1980's an astronomical Search for Extraterrestrial Intelligence be initiated as a long-term effort.

ASTRONOMY AND THE FORCES OF NATURE

Energy Sources in the Universe

In the 1970's, physicists have made substantial progress toward realizing an age-old dream—the understanding of all the forces in nature as different aspects of a single fundamental force. A theory that unifies electromagnetic and weak nuclear forces has been successfully developed along with a comprehensive theory of the strong nuclear force; new theories aimed at unifying both of these theories are now being proposed. Astronomical data have played a role in these developments and may play an even greater role in the future.

Newton's law of gravitation, formulated in precise mathematical terms, set the stage for the investigation of the forces of nature that continues today. We now realize that chemical energy, such as that released in the burning of fossil fuels, is due to the action of electrical forces within atoms. Holding electrons in orbits around nuclei just as gravitation holds planets in their orbits around the Sun, these forces release energy whenever an electron drops into a lower orbit. Magnetic forces result from the motion of electrically charged particles. In the 1860's, Maxwell unified electrical and magnetic forces in a single theory, called electromagnetic theory, which also explains electromagnetic radiation as a wave that sustains itself through a constant interplay between electrical and magnetic energy. By the end of the nineteenth century, both gravitational and electromagnetic forces were well understood at a certain level.

Early in the twentieth century a series of important experiments revealed that the orbits of electrons are qualitatively different from those of planets. The position of a planet can be predicted precisely from a knowledge of the gravitational force acting on it, but the best one can do with an electron is to predict its probability of being at various possible positions. The impossibility of doing any better, embodied in Heisenberg's Uncertainty Principle, is an essential feature of what is now known as quantum theory. Today, the melding of electromagnetic theory and quantum theory, called quantum electrodynamics or QED, is unchallenged in its ability to describe electromagnetic phenomena. A shining goal of contemporary physics is to bring the understanding of all the forces of nature up to the standard of QED.

Sunlight is electromagnetic radiation, and the form in which the energy of sunlight is stored by plants is chemical energy; both forms of energy are embraced by QED. What about the energy stored in the Sun, which it emits as sunlight? Early suggestions included electromagnetic radiation trapped within the Sun, chemical energy stored in its atoms and molecules, and the energy due to the gravitational attraction between all of its atoms. However, none of these forms of energy is adequate to keep the Sun shining for its known age of 4.5 billion years. The solution to this problem was reached in the early 1920's, when it was recognized that a new form of energy discovered in the laboratory, nuclear energy—which is released, for example, when the nuclear force between hydrogen nuclei (protons) draws them together to form helium nuclei—could keep the Sun shining for many billions of years.

Nuclear interactions come into play only at very high temperatures; only then do nuclei have sufficient speeds to overcome their mutual electrical repulsion. Thus, nuclear forces play a role in astronomy only where matter is extremely hot, as in the interiors of stars or in the searing heat of the big-bang explosion. Laboratory studies of nuclear reactions show that there are actually two types of nuclear force, strong and weak; the latter is associated with an unusual particle called the neutrino.

Two Puzzles: Solar Neutrinos and Hidden Mass

Neutrinos can penetrate the entire Sun, so weak is the force with which they interact with matter. Detectors placed beside nuclear reactors, which are copious sources of neutrinos, can record only a minute fraction of those emitted. Despite the great difficulty of detecting them, the role of neutrinos in astronomical research has become increasingly important.

The current theory of stellar energy generation predicts that large numbers of neutrinos are produced in the fusion of hydrogen to helium in the deep interior of the Sun. Because this theory is critical to our understanding of stellar structure and evolution generally, it is important to test this prediction by measuring the flux of solar neutrinos at the Earth. The observed flux of neutrinos is less than one third of that predicted from the most carefully constructed models of the solar interior. Among various proposed explanations of this discrepancy is the possibility that neutrinos behave differently from what has been assumed until recently.

In a completely different area of research, it has been proposed that the problem of hidden mass in galaxies might be resolved if the rest mass of neutrinos were not zero, as usually assumed. From calculations of the number of neutrinos produced in the big bang, one finds that neutrinos could supply the hidden mass in galaxy clusters if they possess a rest mass about 1/10,000 that of the electron.

There are thus two astronomical problems that might be resolved if neutrinos prove to have properties not previously known. Theoretical physicists have recently suggested a resolution of both of these problems. The recently developed unified theory of weak and electromagnetic forces is based on a principle called gauge invariance and is therefore referred to as "the gauge theory of weak and electromagnetic interactions." So far it has succeeded in explaining all the various phenomena involved with both electromagnetic and weak nuclear forces.

The gauge theory of weak and electromagnetic interactions in its original form says nothing about the problems of solar neutrinos or hidden mass. However, pursuing the principle of gauge invariance behind it, physicists have constructed a theory of the strong nuclear force, called quantum chromodynamics, or QCD. This theory postulates the existence of elementary particles that combine to form protons and neutrons, called quarks. The success of QCD in explaining the results of experiments in elementary-particle physics gives increasing confidence that it is the correct theory of the strong force that binds neutrons and protons into atomic nuclei.

Spurred by the success of the gauge theory of weak and electromagnetic interactions and of QCD, physicists are now trying to find an even more general gauge theory, called "Grand Unified Theory," that incorporates both. Some theories of this type predict that there should be the three types of electrons that are actually observed, as well as three corresponding types of neutrinos, called e, mu, and tau. In some versions of the theory, e, mu, and tau neutrinos are regarded as three aspects of the same basic neutrino, which has a finite rest mass and which oscillates back and forth among its three aspects. Although the nuclear reactions in the Sun emit only e neutrinos, according to some Grand Unified Theories neutrino oscillations would be expected to occur long before the neutrinos reached the Earth, so that at the Earth one would observe a random mixture of e, mu, and tau neutrinos. Since the Homestake Mine apparatus is sensitive only to e neutrinos, a factor-of-3 discrepancy would thereby be explained.

Oscillations can occur only if neutrinos have a finite rest mass. If the value of the rest mass were in the right range, it would have a dramatic bearing on our understanding of the hidden-mass problem and of the ultimate fate of the Universe.

Theories involving several different types of neutrinos are constrained by calculations of the properties of the early Universe. If there were more than about four types of neutrinos, their contribution to the gravitational acceleration in the early Universe would have been so great that there would not have been sufficient time for primordial neutrons to decay; there would then be more helium in the Universe than is actually observed. Thus, current astronomical observations eliminate some versions of Grand Unified Theories.

A critical experiment endorsed earlier in this report will help to shed light on the true nature of neutrinos. The gallium solar neutrino experiment will be sensitive to neutrinos of much lower energy than those measured by the ^{37}Cl detector in the Homestake Mine. The flux of such lower-energy neutrinos can confidently be calculated from the observed luminosity of the Sun, independently of the details of solar models. If there is a discrepancy between the predicted and observed values of the solar neutrino flux in the gallium experiment, it could be an indication that neutrinos oscillate and have a finite neutrino rest mass.

There may also be powerful sources of high-energy neutrinos among the many sites of violent activity observed to occur on both stellar

Chlorine solar-neutrino detector deep in the Homestake Mine, Lead, South Dakota. (Photo courtesy of R. Davis, Jr., Brookhaven National Laboratory)

and galactic scales. Despite the difficulty of detecting such neutrinos and the weak fluxes to be expected because of the distances to the sources, the study of energetic-neutrino detectors with possible astronomical applications is appropriate for the coming decade. An interesting possibility for such study is the proposed observation of neutrino-induced reactions in seawater employing arrays of photomultipliers to detect the associated Cerenkov radiation.

Before the First Three Minutes

Although astronomical data now available appear to be in agreement with the predictions of big-bang cosmology, the big-bang model cannot yet be considered conclusively proven, so that it is of the greatest importance to test its predictions however possible. In particular, the model predicts that the cosmic microwave background originated as high-temperature radiation in the first few minutes of time. As the Universe expanded, according to this view, the radiation cooled to its present observed temperature, about 3 degrees above absolute zero. When the Universe was about 1/10,000 of a second old, its temperature was a trillion degrees, so hot that the radiation present created about 100 million proton–antiproton pairs for every proton now observed in the Universe. As time passed, these pairs annihilated, leaving behind only the very small fraction by which the number of protons exceeds the number of antiprotons. Had this excess not existed, the number of protons in the present Universe would have been 10 billion times smaller, and there would not have been sufficient matter in the Universe for the formation of galaxies, stars, and planets.

What caused the excess of matter over antimatter implied by this big-bang scenario? Until recently, physicists had regarded the excess as a fact as inexplicable as the existence of the Universe itself. Recently, it has been suggested that Grand Unified Theories provide an explanation: very heavy particles present in the first 10^{-38} sec of the history of the Universe decayed, creating in the process slightly greater numbers of protons than antiprotons. This prediction can be tested in a straightforward way, for if protons can be created they must also decay. As the lifetime of the proton estimated from Grand Unified Theory is 100 billion billion times the age of the Universe, physicists are not concerned that the Universe will soon evaporate. On the other hand, the predicted proton lifetime is sufficiently short that one such decay will occur in a ton of material each year. Experiments are now in progress to detect such events.

The Limits Of Gravitation

Gravity keeps us on the Earth, binds the Earth to the Sun, and slows the expansion of the Universe. Newton described it as a force, while Einstein, in his General Theory of Relativity, interpreted gravitational forces in terms of the curvature of space–time. Einstein's theory, unlike Newton's, is believed to be valid for very strong gravitational fields and for bodies moving close to the speed of light; it is therefore crucial for an understanding of systems such as neutron stars, black holes, and the expanding Universe.

The General Theory of Relativity predicts that when any non-spherical body collapses to form a compact object or a black hole, it emits a new form of energy called gravitational radiation. Although this radiation is predicted to be extremely difficult to detect, several research groups are now building detectors thousands of times more sensitive than those available during the 1970's. Parallel efforts to calculate the amount of gravitational radiation emitted by collapse indicate that, if the planned development of new instrument concepts succeeds, we might hope to detect an event within two decades—

even earlier if there should be a new supernova within the Galaxy. The recently confirmed, slow decrease in the orbital period of the binary pulsar has already been interpreted as the result of gravitational radiation from a close pair of neutron stars.

While efforts to develop a quantum theory of gravitation have not yet succeeded, there is reason to believe that quantum effects should occur near black holes, where space–time curvature is high. The quantum theory of elementary particles predicts that even in vacuum, particle–antiparticle pairs are constantly being produced and annihilated in an interval of time too short to observe. If this effect should occur near a black hole, one member of the pair may fall into the black hole before the pair annihilates. Zero-mass particles, including photons, are created similarly; the black hole thus appears to the outside world as a source of radiation, ultimately evaporating as a result of the energy lost. Black holes of all sizes could have been created in the big bang; in particular, those having masses about of 10^{15} g (the mass of a small mountain on Earth) would just be evaporating now, giving rise at the ends of their lives to bursts of gamma radiation. Such radiation from evaporating black holes has been searched for, and, although the Gamma Ray Observatory will continue the quest, so far none has been found. It thus appears that primordial black holes with masses less than that of a mountain cannot make up a significant fraction of the mass of the Universe.

The theory of black-hole evaporation depends on the quantum nature of strong nuclear forces but not on the quantum nature of gravitation. Although no convincing theory of gravitation that incorporates the quantum principle has yet been produced, it is conjectured that the quantum effects must become important whenever the radius of curvature of space–time becomes less than the so-called Planck length, 10^{-33} cm. Such conditions are thought to have occurred in the Universe at times before 10^{-43} sec and at temperatures above 10^{32} deg. Because the energies and temperatures characteristic of Grand Unified Theories are remarkably close to these values, some physicists believe that a theory should be possible that incorporates all four forces in nature into one "Super-Grand" force at energies only slightly higher than those relevant to Grand Unification. A prime hope for such a theory is that it will yield, almost as a by-product, the correct theory of quantum gravitation. Attempts in this direction have so far met with little if any success, but the development of such a theory could be considered to be the ultimate challenge to physics at present.

The notion of force, as a law governing matter once created, fails to take account of the process of creation itself. Is it possible, as astrophysics pushes the frontiers of time back to the moment of cosmic creation, that the existence of the Universe will be recognized as a consequence of the nature of the fundamental force? Is it possible that the potential existence of the world somehow calls it into existence? Such questions, once believed outside the range of science, are now arising in scientific thought.

The primary mirror for the Space Telescope being inspected after figuring. Photo courtesy of the National Aeronautics and Space Administration)

4

Approved, Continuing, and Previously Recommended Programs

As emphasized in Chapter 2, the achievement of the goals for astronomical research presented in Chapter 3 depends heavily on the maintenance of support for approved and continuing programs; these, together with programs previously recommended in other National Academy of Sciences reports for implementation in fiscal year 1982 and earlier, form the base of present and planned resources from which the recommendations of the Astronomy Survey Committee proceed. Their role in the research of the 1980's is discussed below in the following order, which carries no implication of priority:

A. Space Telescope and the associated Space Telescope Science Institute;

B. Second-generation instrumentation for Space Telescope;

C. The Gamma Ray Observatory;

D. Level-of-effort observational programs within the National Aeronautics and Space Administration:
- The NASA Explorer satellite program, with a substantial augmentation,
- Research with balloons, aircraft, and sounding rockets, at enhanced levels of support, and
- The Spacelab program, reaffirming NASA's original strong commitment to research with the Space Shuttle;

E. Two major astrophysics facilities for Spacelab:
- The Shuttle Infrared Telescope Facility (SIRTF) and

- The Solar Optical Telescope (SOT);

F. Facilities for the detection of solar neutrinos;

G. Federal grants in support of basic astronomical research at U.S. universities;

H. Programs at the National Astronomy Centers; and

I. The 25-Meter Millimeter-Wave Radio Telescope.

The remainder of the present chapter illustrates the role that these programs will play in addressing the major scientific problems of the coming decade.

A. SPACE TELESCOPE AND THE ASSOCIATED SPACE TELESCOPE SCIENCE INSTITUTE

The Astronomy Survey Committee regards Space Telescope (ST) as a project of extreme importance for all of astronomy and recommends that NASA complete its development and place it into operation at the earliest possible date. This facility will represent one of the most momentous advances in astronomical instrumentation since Galileo's first telescope.

The launch of ST, now planned for 1985, will provide the first permanent optical observatory in space. This facility, carrying a 2.4-m telescope of superb optical quality, will take full advantage of the benefits of observing above the Earth's atmosphere: sharp images unaffected by clouds or atmospheric turbulence, elimination of airglow, and extension of the spectral range into the ultraviolet (UV) and near-infrared wavelength regions. ST will be the first orbiting telescope large enough to carry out studies of extragalactic objects at the limits of the observable Universe. The first complement of instruments on ST will obtain digital photometric images over the wavelength range 1200–12000 Å, spectra from 1200 to 8000 Å with a large range of resolutions, and visible and UV photometry with fast time resolution. The European Space Agency will provide one of these instruments as part of a planned international collaboration in the ST program.

The results of ST will profoundly affect every branch of astronomy. Identification and photometry of stars fainter than the twenty-seventh magnitude in the visible and UV regions will revolutionize extragalactic astronomy. Studies of stellar populations down to the main sequence can be made in nearby galaxies and in almost all of the globular clusters of our Galaxy. Bright stars suitable as distance indicators can be observed individually in galaxies out to the Virgo cluster, allowing the cosmic expansion to be studied with unprecedented accuracy. The high angular resolution of ST will permit morphological classification of galaxies with red shifts up to order unity, allowing direct studies of the evolution of galaxies over the past 5 billion to 8 billion years. What kinds of galaxies may be associated with quasars is likely to be determined from an analysis of ST images. The question of whether some galaxies harbor giant black holes in their nuclei may be settled. Galactic astronomy will utilize the ability to measure the very faint stars in globular clusters and the Galactic bulge to determine, for example, the mass function of star formation in various regions of the Galaxy. The imaging capability of ST will also open an important new range of investigations within the solar system; for example, diffraction-limited images at high spatial resolution will permit detailed views of the structure and dynamics of atmospheric circulation on other planets.

The spectroscopic capabilities of ST are equally impressive. Our understanding of the interstellar medium, revolutionized by observations with the *Copernicus* satellite and rocket surveys of soft x rays, will be further advanced by ST measurements down to 1200-Å wavelength: new observations will be made at much higher spectral resolution and in much more heavily obscured regions all over the Galaxy. Most of the known quasars can be studied spectroscopically in the UV region, permitting critical comparisons of distant and nearby objects and the use of quasars as probes of the intervening medium. In planetary astronomy, the UV spectroscopic capability will permit searches for new molecules in planetary atmospheres and comets and delineate their structure. The surface chemistry of planets and asteroids can also be explored.

The Committee strongly approves NASA's commitment to operate and support ST through an associated but independent Space Telescope Science Institute, which will include international participation. The Institute will be responsible for the scientific direction of ST, for data-reduction facilities, for education concerning use of the instrumentation, and for meeting all research expenses incurred by the ST user community; it will also be the liaison between NASA and the scientific community on matters concerning the management and improvement of this powerful instrument.

B. SECOND-GENERATION INSTRUMENTATION FOR SPACE TELESCOPE

Space Telescope is designed to employ interchangeable instruments that can be replaced either in orbit or when ST is returned to Earth. Since ST should exploit all the benefits of the latest technology, the Astronomy Survey Committee regards timely upgrading as extremely important. Major changes that can currently be foreseen include improvements to the ST spectrographs and the implementation of ST's potential infrared capability. These two changes are sufficiently important to merit further discussion.

Both of the first-generation spectrographs on ST employ Digicon detectors with one-dimensional diode arrays behind a photocathode. The power of a spectrograph can be increased dramatically through use of a two-dimensional detector; in an echelle format, for example, a wide spectral range is then available in one high-resolution exposure; for extended objects, spatial information is then obtained along the entire slit image. It seems likely that two-dimensional, charge-coupled-device-type detectors with very low read-out noise and high UV sensitivity will become available within a few years. This combination of two-dimensional coverage with high quantum efficiency constitutes an upgrading of very high priority.

Present ST instruments do not allow observations at wavelengths much longer than 1 μm. Although the infrared region will also be covered by the IRAS Explorer satellite and the powerful SIRTF Shuttle facility, the large aperture of ST gives it a great advantage in two crucial areas of scientific investigation. One is imagery in the near infrared, for which detectors of high quantum efficiency will almost certainly soon become available. The other is high-resolution spectroscopy throughout those extensive regions of the infrared to which the Earth's atmosphere is opaque; this work requires the greater collecting area and angular resolution of ST. Infrared capability in each of these areas is complementary to both ground- and space-based programs and will be valuable for a wide variety of critical programs ranging from planetary science to cosmology. For example, since the expansion of the Universe shifts the well-studied optical region into the infrared for very distant objects, evolutionary studies of galaxies cannot be complete without infrared data. The Committee therefore recommends that NASA develop appropriate infrared instruments to be flown aboard ST.

C. THE GAMMA RAY OBSERVATORY

The Astronomy Survey Committee believes that the Gamma Ray Observatory (GRO) will yield results on high-energy processes that will be of fundamental importance to the advance of astrophysics during the coming decade. The 1979 report of the Space Science Board's Committee on Space Astronomy and Astrophysics (CSAA) recommended that GRO be the next new space-astronomy mission beyond those already approved by Congress at that time (*A Strategy for Space Astronomy and Astrophysics for the 1980's*, National Academy of Sciences, Washington, D.C., 1979). The Survey Committee joins with CSAA in endorsing this important facility and is pleased that GRO is now an approved component of the U.S. space-science program.

Gamma-ray astronomy permits the study of energy transformations in critically important processes, such as cosmic explosions, acceleration and interactions of high-energy particles, gravitational accretion by superdense objects, nucleosynthesis in stars, and matter–antimatter annihilation. Because gamma rays have high penetrating power, they can reach the Earth from parts of the Universe whose optical or low-energy x-ray emission may be obscured by intervening matter, such as the center of our own Galaxy and the central regions of active galaxies.

Observations with GRO will address many important astronomical questions. The SAS-2 satellite and, in Europe, the COS-B satellite have demonstrated the rich character of the diffuse Galactic radiation, believed to be a result of cosmic-ray interactions with the interstellar medium. Observations of this emission on a finer scale with GRO will help to determine the origin and dynamic-pressure effects of cosmic rays, to identify large concentrations of interstellar gas, and to study Galactic structure.

Numbers of discrete gamma-ray sources have been detected in previous surveys. While some of these are associated with known supernova remnants and pulsars, others have not yet been identified at longer wavelengths and may thus represent a new class of objects. Additional gamma-ray observations are required to clarify why some sources are more complex than anticipated. The Vela pulsar, for example, exhibits two gamma-ray pulses per cycle, while only one pulse occurs in the radio region; moreover, neither gamma-ray pulse is in phase with the radio pulse. Another puzzle is a strong source in the general direction of the Galactic center, which is indicative of pair annihilation; further observations are needed to investigate the nature of this source.

GRO can obtain information relevant to nucleosynthesis by detecting nuclear-decay gamma-ray lines of recently synthesized elements in our own and other nearby galaxies. In the most intense sources it may be possible to record line profiles, from which one can infer the current rate of heavy-element production and information relevant to the densities, temperatures, and flow velocities in supernova remnants. Regions in which massive stars are forming may be found through observations of the lines of isotopes such as ^{26}Al.

GRO will permit the study of cosmic gamma-ray bursts with much greater sensitivity and energy resolution than provided by any previous or other currently planned mission, thus shedding new light upon the nature of gamma-ray burst sources.

GRO will also obtain detailed information on the spatial uniformity and energy spectrum of the diffuse extragalactic gamma radiation, providing clues to its origin. Since gamma radiation may reflect most directly the primary emission process in active galaxies, GRO observations will contribute to an improved understanding of these objects. Studies of time variability at gamma-ray (and other) wavelengths will yield insight into the active volumes and associated energy densities. High-energy gamma rays may also be observed from nearby normal galaxies; if so, GRO data will furnish information on the distributions of cosmic rays and matter in such galaxies as well.

The Committee agrees with CSAA that the continued development and timely launch of GRO is essential to the pursuit of these important scientific objectives during the coming decade.

D. LEVEL-OF-EFFORT OBSERVATIONAL PROGRAMS WITHIN THE NATIONAL AERONAUTICS AND SPACE ADMINISTRATION

NASA has developed three modes for carrying out observational space astronomy within level-of-effort programs: the Explorer program; balloons, aircraft, and sounding rockets; and the Spacelab program. Each of these modes provides a highly effective and in some cases unique mechanism to address particular observational problems in astronomy and to test instrument concepts employing state-of-the-art technology. The Astronomy Survey Committee regards these level-of-effort programs as the backbone of observational space astronomy and astrophysics for the 1980's. The Committee therefore urges the continued, vigorous support of the Explorer program and of research with balloons, aircraft, and rockets, together with a speedy reaffirmation of NASA's original strong commitment to the Spacelab program.

The Explorer Program

The Explorer program has been one of the most productive and cost-effective elements of the NASA space-science program. Its level-of-effort character has encouraged both frugal management and relatively rapid response to newly perceived scientific opportunities, providing a highly effective mechanism to exploit new observational techniques, to explore newly accessible wavelength intervals, or to study a particular class of objects. Nearly all of the branches of space astronomy have had or will soon have their beginnings in Explorer missions. For example, the present International Ultraviolet Explorer (IUE) program has provided essentially all of the intermediate-resolution UV spectra available to the world of astronomy. The future astronomy Explorers now under development or planned give promise of continuing this high level of pioneering achievement:

• The Infrared Astronomy Satellite (IRAS), a project based on international collaboration, will provide a detailed and comprehensive reconnaissance of those components of the Universe that radiate most strongly at relatively low temperatures (100–1000 K). Millions of sources that emit substantially in the 10–100-μm wavelength range should be detected and their positions determined by IRAS, providing a wealth of material for an initial survey of regions of active star formation as well as of the brightest extragalactic sources. IRAS will also be able to determine the surface composition of thousands of asteroids within the solar system.

• The Cosmic Background Explorer (COBE) will make definitive measurements of the 3 K cosmic background radiation, now generally accepted to be a relic of the radiation generated in the big bang. COBE will test this conclusion through precise measurements of the intensity throughout the spectrum and of the anisotropy of the radiation. Deviations from isotropy on the largest angular scales reflect the motion of our Galaxy with respect to the large-scale structure of the

Universe, while deviations on angular scales of the order of 10° carry unique information about the inhomogeneity of the early Universe.

• The Extreme Ultraviolet Explorer (EUVE) will open for study another major region of the electromagnetic spectrum, from 100 to 912 Å wavelength (from soft x rays to the UV region below the Lyman limit), a primary objective being the completion of an unbiased, all-sky survey of sources of EUV radiation. EUVE is expected to reveal a large number of new sources with temperatures in the range 10^5–10^6 K, furnish broadband spectral information on many other already known sources, and provide new information on the structure and ionization state of the interstellar medium over a wide range of distances from the Sun. This mission can also yield new information about the Jovian magnetosphere and cometary atmospheres.

• The X-Ray Timing Explorer (XTE) will provide important new opportunities for observations of variability in x-ray sources on time scales ranging from milliseconds to years. The scientific objectives of this mission include investigations of the mass, magnetic moment, and internal structure of neutron stars and degenerate dwarfs; the physics of accretion disks, plasmas, and stellar magnetospheres; the geometry of source emission regions; the nature and evolution of normal stars, through studies of mass loss; the nature of variable sources, such as x-ray bursters and transient x-ray sources; and the underlying physics and emission mechanisms in compact extragalactic objects.

The Committee endorses the above mission concepts, which have been carefully studied and highly recommended by other review groups. Our concern is primarily for the future of this highly productive program. We believe that it is vitally important for NASA to maintain a strong Explorer program; over the past decade, however, inflation and other factors have doubled the dollar cost charged to the Explorer budget for the same real level of effort. The Committee therefore recommends an augmentation to the Explorer program to restore it at least to the level of effort of the previous decade. Such an augmentation will ensure not only the timely flights of the missions described above but also opportunities for pursuit of many other exciting new scientific objectives, as discussed in Chapter 6.

Balloons, Aircraft, and Sounding Rockets

The balloon, aircraft, and sounding-rocket programs have been vitally important to the progress of astronomy and to the development and testing of new detector systems later used on satellites. The Greenstein report recommended a doubling of support for balloons, rockets, and aircraft; balloon research has received strong support by the Balloon Study Committee of the Geophysics Research Board (*The Use of Balloons for Physics and Astronomy*, National Academy of Sciences, Washington, D.C., 1976); and balloons, aircraft, and rockets have been supported by CSAA in its recent report (*A Strategy for Space Astronomy and Astrophysics for the 1980's*, National Academy of Sciences, Washington, D.C., 1979). Unfortunately, funding for such programs has not grown, although the need is ever more acute. The present Committee supports a vigorous effort in all three areas during the 1980's to encourage innovative experiments and to obtain promptly the initial scientific results from recently developed instruments.

The balloon program has been of particular value to infrared, x-ray, gamma-ray, and cosmic-ray astrophysics, yielding important contributions to the study of cosmic rays, energetic x-ray spectra, low-energy gamma-ray bursts, gamma radiation from the Galactic center, and far-infrared emission from ionized H II regions. Balloon detector systems were prototypes for instruments on HEAO-3 and SAS-2 and for several of those planned for COBE, Spacelab, and GRO. The balloon program will continue to provide important new scientific results and permit new instruments to be tested in an environment similar to that of space. In order to remain productive, this program should receive an augmentation in funding to compensate for inflation, to provide larger balloons for heavier instruments, and to develop a new balloon system to allow flights of longer duration.

The NASA aircraft program—including the U2 aircraft, the Lear Jet Observatory, and especially the Kuiper Airborne Observatory (KAO)—has achieved scientific results of great importance, including the first observations of far-infrared emission from other galaxies; an important series of infrared observations of our own Galactic center; the study of internal energy sources in Jupiter, Saturn, and Neptune; the probing of interstellar molecular clouds and ionized regions in new ways; a primary role in the discovery of rings around Uranus; and studies of water vapor in the atmosphere of Jupiter and of sulfuric acid droplets in the clouds of Venus. The KAO and Lear Jet Observatory not only continue to serve as test facilities for many of the research instruments and techniques being developed for space-science applications but also provide the reconnaissance of the field necessary for future major missions. For example, first-generation instrumentation for far-infrared and submillimeter-wavelength spectroscopy is now being used on KAO to probe the spectral lines that are believed to govern the energy balance in the bulk of the interstellar medium. The full potential of these aircraft observatories unfortunately has not been achieved because of a lack of funds for operations and personnel. A substantial increase in funding should be provided for KAO operations in particular, not only to compensate for increased fuel costs but also to permit the greater numbers of flights required for a more intensive and hence more efficient use of this outstanding facility.

The U.S. sounding-rocket program has played a key role in advancing the frontiers of x-ray, ultraviolet, and infrared astronomy, with achievements including the discovery of the first cosmic x-ray source (Sco X-1), discovery of x-ray pulsations from the Crab pulsar, the first measurement of the UV spectrum of a nearby quasar, the first soft x-ray sky survey, and the first 5–20-μm all-sky survey, which yielded new information about classes of previously unobserved infrared sources. Sounding rockets will continue to provide economical and effective means for developing new instruments, for testing new observational techniques, and for exploratory investigations. A strong sounding-rocket program should therefore be maintained. The Committee also endorses the extension of observing time for rocket payloads through their placement in temporary orbit by the Space Shuttle as part of NASA's Experiment of Opportunity Program (EOP).

The Spacelab Program

The Spacelab program will provide new flight opportunities for large payloads that require servicing and will facilitate observational programs demanding higher altitudes or flights of longer duration than can be achieved by aircraft or balloons. Spacelab flights will be particularly suitable for those facilities (both large and small) that can gather substantial quantities of data within the relatively short initial flights of the Space Shuttle; the Solar Optical Telescope and Shuttle Infrared Telescope Facility described in Section E below are examples. In its recent report, cited in the preceding section, CSAA recommended a vigorous program of astronomy and astrophysics on Spacelab with an annual level of effort exceeding that of a typical moderate-class mission.

It is therefore of serious concern that funding for Spacelab experiments and facilities has not yet reached substantial levels; support for experiments that have already been approved has been significantly reduced, and the selection of additional PI experiments has been deferred. The Committee is cognizant of the programmatic and technical problems that have contributed to these decisions. However, now that flights of the Space Shuttle have begun, we urge NASA to re-establish its original strong commitment to a vigorous Spacelab program as soon as possible.

The Spacelab program will also facilitate the development of large instruments designed to make initial observations on Shuttle flights and later to be placed in orbit to carry out more comprehensive scientific programs, possibly aboard a long-duration space platform. The augmentation of Spacelab capability by such a space platform would allow the Shuttle to realize its full potential as a scientific tool. We therefore support NASA plans for the development of a long-duration space platform operated in conjunction with Spacelab and the Shuttle (see Appendix A, Statement Concerning a Space Platform). We furthermore endorse the pursuit of means to extend the observing time available to rocket-sized payloads through the development of a standard module that will allow such payloads to be placed in temporary orbit during Spacelab missions, without incurring the substantial costs of an active Shuttle interface, such as may be required for larger payloads.

The Committee also encourages the development of a Solar Shuttle Facility, composed of several advanced facility-class instruments, to be used in a coordinated group on the Space Shuttle. These instruments will obtain data critical to an understanding of the fundamental plasma processes underlying cyclic activity and transient high-energy phenomena on the Sun and other stars; their development should proceed as part of the ongoing Spacelab program, with the Solar Optical Telescope (SOT) as the first such instrument. A plan for the addition of succeeding instruments is expected to emerge from the recommendations of other advisory groups, particularly the CSSP. Among the other instruments proposed for inclusion are a Solar Soft X-Ray Telescope Facility (SSXTF), a Grazing Incidence Solar Telescope (GRIST, selected for advanced study by the European Space Agency), and a Pinhole/Occulter Facility for hard x-ray imaging and for the study of the corona at high resolution (currently under study by NASA). Still other solar instruments—such as those for EUV and gamma-ray observations and for specialized observations of the extended corona, long-period photospheric oscillations, and large-scale circulation—should be considered for development as PI instruments or as additional facilities within the Spacelab program.

E. TWO MAJOR ASTROPHYSICS FACILITIES FOR SPACELAB

The Astronomy Survey Committee endorses both the Shuttle Infrared Telescope Facility (SIRTF) and the Solar Optical Telescope (SOT) as the first major astrophysics facilities planned for Spacelab; the order of the following discussion carries no implication of priority.

The Shuttle Infrared Telescope Facility (SIRTF)

The proposed SIRTF will be the cornerstone of research in infrared astronomy during the 1980's. The Astronomy Survey Committee joins with the Space Science Board's Committee on Space Astronomy And Astrophysics (*A Strategy for Space Astronomy and Astrophysics for the 1980's*, National Academy of Sciences, Washington, D.C., 1979) in recommending this facility as the first major infrared telescope in space.

SIRTF will permit investigations over the enormous range of wavelengths from 2 to 300 μm. For some important types of observations it will, because of its cryogenically cooled optics, yield a sensitivity gain of 1000 over the largest existing ground-based and airborne infrared telescopes; this gain in sensitivity is so large that it is not unreasonable to expect that SIRTF will make important and unexpected discoveries. The multiple, interchangeable focal-plane instruments planned for SIRTF will moreover greatly increase our ability to explore the evolution of distant extragalactic sources, the physical properties and chemical composition of molecular clouds and regions of star formation, the nature of cometary nuclei and asteroids, and the structure of planetary atmospheres. For example, SIRTF will be able to detect infrared sources at the limits of the observable Universe; on one Shuttle flight, it could gather information on sources of both large and small red shift, thus permitting a comparison of the energetics of quasars and galaxies at the earliest epochs of the Universe with those at the present epoch. Because of its relatively wide field of view, SIRTF will also be able to carry out efficient surveys of infrared sources that will help to optimize the observing programs of larger instruments, such as the New Technology Telescope (NTT), the VLB Array, and a Large Deployable Reflector (LDR) in space, which have narrower fields of view.

SIRTF should be an early and frequently flown payload on Shuttle sortie missions. It has such high sensitivity that very extensive infrared observations can be accomplished even within the relatively brief, 7-day profile of an early Shuttle flight; reflight on the Shuttle makes it possible to fly SIRTF with continually improved focal-plane instrumentation and detectors. Eventually, however, SIRTF flights of longer duration will be needed to realize the full potential of this remarkable facility, and the study of such flights is recommended in Chapter 7.

The Solar Optical Telescope (SOT)

The SOT will carry out solar observations from 1100-Å wavelength into the near-infrared wavelength region with high angular resolution—0.1 arcsec at 5000 Å (70 km on the Sun) and nearly 0.02 arcsec at 1100 Å (14 km)—together with very high spectral resolution. Following the recommendations of the Space Science Board's Committee on Space Astronomy and Astrophysics (*A Strategy for Space Astronomy and Astrophysics for the 1980's*, National Academy of Sciences, Washington, D.C., 1979) and Committee on Solar and Space Physics (*Solar-System Space Physics in the 1980's: A Research Strategy*, National Academy of Sciences, Washington, D.C., 1979), NASA has selected SOT as the first major astrophysics facility to be developed within the Spacelab program.

Solar observations from the ground and from space have demonstrated that fundamental processes governing solar phenomena occur on size scales smaller than can be resolved by present instruments (several hundred kilometers). SOT will achieve a tenfold improvement in spatial resolution for the study of these processes. It will also furnish the stability and spectral resolution required for sensitive measurements of radial bulk motions and oscillations of the solar atmosphere, important for an understanding of the convective processes that control magnetic and atmospheric structure and the 11-year cycle of sunspots and solar magnetic activity.

SOT will also measure flare phenomena with unprecedented precision. Numerous ionic emission lines characteristic of flare-produced plasmas fall within the spectral range of SOT; measurements of these lines will provide a unique picture of the energy-release process that is believed to heat the chromospheric plasma to x-ray incandescence. Such observations may, for the first time, allow the metastable magnetic-field configurations that precede a flare to be determined with a spatial resolution approaching 20 km.

As new focal-plane instruments are developed and installed for later flights, the scientific impact of SOT will broaden to encompass most other major areas of solar research, including the structure and dynamics of the solar convection zone; the dynamo responsible for the solar magnetic field and the solar activity cycle; and the thermodynamic structure and dynamical behavior of the solar photosphere, chromosphere, transition region, and corona. A more complete understanding of these solar properties should also lead to a better understanding of physical processes in high-temperature plasmas, energy-release mechanisms in other objects such as quasars, and activity cycles in other stars.

F. FACILITIES FOR THE DETECTION OF SOLAR NEUTRINOS

The Astronomy Survey Committee recommends continued, vigorous support for programs to detect and measure the flux of neutrinos from the interior of the Sun. Additional facilities are needed to supplement the data currently being obtained by ^{37}Cl detectors at underground sites, and the use of a neutrino detector employing substantial quantities of gallium presents a particularly attractive opportunity to broaden our knowledge of the solar-neutrino energy spectrum. Detectors employing ^{7}Li and ^{115}In may also become feasible during the coming decade.

The observation of neutrinos produced in the interior of the Sun is our only direct source of information about the process of stellar energy generation, which is of fundamental importance to an understanding of the structure and evolution of the Sun and other stars. Experiments of the past decade employing ^{37}Cl detectors at underground sites suggest strongly that relatively high-energy neutrinos from the solar interior are arriving at the Earth at no more than one third of the rate predicted by theory. Although this finding has prompted searching re-examination of the theory of stellar energy generation, of currently accepted models for solar structure, and of neutrino physics itself, no generally accepted explanation for the discrepancy has yet been found. The revision of present concepts in any one of these three areas of inquiry would have profound implications for astronomy and astrophysics.

It is important that the ^{37}Cl experiments be continued and refined so that they may yield more precise data. At the same time, it is essential to pursue the solar-neutrino problem with additional, complementary experiments aimed at recording those neutrinos produced in other chains of nuclear reactions inside the Sun. The use of substantial quantities of gallium as a detector offers an outstanding opportunity in this regard and permits the concurrent test of relevant neutrino properties as well. For a conclusive measurement of the solar-neutrino flux in the coming decade, approximately 50 tons of gallium will be required; however, the gallium will not be consumed during the experiment envisaged and may be sold at its completion.

A pilot experiment employing 1.3 tons of gallium was successfully completed during the summer of 1980. The U.S. Department of Energy and the Max Planck Institutes in Germany are now supporting the development of a larger detector that will be calibrated with a laboratory source of neutrons produced in a reactor. The Committee urges continued financial support of the gallium-detector experiment by this international collaboration and the completion and operation of the full gallium neutrino detector at the earliest opportunity.

G. FEDERAL GRANTS IN SUPPORT OF BASIC ASTRONOMICAL RESEARCH AT U.S. UNIVERSITIES

The Astronomy Survey Committee recommends the continuation of programs of federal grants in support of basic astronomical research at U.S. universities. The most important of these are the grants program of NSF's Astronomy Division and grants awarded through NASA's Research and Analysis Program (formerly the Supporting Research and Technology Program); the research needs of the 1980's will require substantial increases in both of these funding sources.

U.S. astronomical research is carried out in a number of different types of organizations, including federal laboratories, the National Astronomy Centers, privately endowed research institutes, private industry, and private and state universities. The diversity of these organizations is one of the strengths of the national research effort in astronomy; for example, U.S. research in radio, x-ray, and UV astronomy was initiated at federal and industrial laboratories, while optical astronomy started much earlier in universities and research institutes.

Vigorous programs of basic astronomical research at U.S. universities are of special importance in this overall effort because universities are responsible for the training of future research scientists and teachers of astronomy and, through their teaching programs, for the dissemination of the latest scientific results to students and the general public. In this process, the interaction of established astronomers with bright young people stimulates many new ideas for research and for the development of new techniques. The great majority of visiting astronomers who use the National Astronomy Centers are university scientists; the overall scientific productivity of the Centers is thus itself heavily dependent on the quality of the university research environment. The strength and vigor of university research in turn depends decisively on federal support through grants to university scientists.

Federal support of university research in astronomy flows from a number of sources. The most broadly distributed and critically needed support of university research comes from grants awarded through NASA's Research and Analysis (R&A) Program, which is essential to the health of U.S. space astronomy, and from the grants program of NSF's Astronomy Division, which provides support to greater numbers of U.S. university astronomers than any other federal funding source.

NASA's R&A program has been highly effective in supporting basic research of importance to the planning of space-science missions. Those fields of astronomy that are restricted to observations from balloons, rockets, aircraft, or space vehicles—such as gamma-ray, x-ray, and UV astronomy, together with cosmic-ray studies—rely almost totally on R&A funding for initial research efforts. The R&A program has been critical for the development of instrumentation, detectors, and other hardware for space astronomy and will continue to play this key role throughout the coming decade; R&A support of theoretical astrophysics should also increase substantially in the years ahead.

NSF's mission to support U.S. ground-based astronomy finds a particularly effective expression at the universities through the Astronomy Division grants program, which is the primary source of operating funds for many of the nation's major university-operated radio-astronomy observatories and which also provides extensive support for instrumentation at many ground-based optical observatories operated by universities. Most theoretical astrophysics in the United States is also supported through the NSF grants program (with a small but increasingly important contribution by NASA). The purchase of computers, the development of detectors, the construction of instrumentation, and the support of technical personnel are other examples of the critically important needs met by NSF grants program.

Finally, both the R&A program of NASA and the NSF grants

program are essential for the effectiveness of university research activities, providing funds for travel, publication costs, and computation, together with student, postdoctoral, and—in the summer—faculty salaries. These funds, which are awarded on the basis of merit, reach an extremely wide range of the U.S. astronomical community and foster a cooperative spirit that benefits all involved. The universities provide the research environment and academic-year salaries to scientists; NASA and NSF provide these scientists with the appropriate tools for research.

For these reasons, the NASA and NSF grants programs will continue to be essential to the health of basic astronomical research at U.S. universities in the coming decade. Substantial increases above present funding levels in both programs will be necessary to provide for expanded needs for instrumentation and detectors, theory and data analysis, computational facilities, laboratory astrophysics, and technical support at ground-based observatories, as discussed in Chapter 5.

H. PROGRAMS AT THE NATIONAL ASTRONOMY CENTERS

The Astronomy Survey Committee recommends the vigorous support of programs carried out at the National Astronomy Centers. Five of the Centers receive support from NSF's Astronomy Division:

- Sacramento Peak Observatory, one of the nation's leading institutions for ground-based observational research in solar physics;
- Kitt Peak National Observatory, operator of the most heavily utilized and broadly instrumented collection of optical and infrared telescopes generally available to American astronomers;
- Cerro Tololo Inter-American Observatory in Chile, a vital optical/infrared window on the southern hemisphere of the sky, operated with the cooperation of Chilean astronomers, who share in the observing time;
- National Astronomy and Ionosphere Center, operator of the world's largest (300-m) single-dish radio telescope at Arecibo, Puerto Rico; and
- National Radio Astronomy Observatory, operator of the nation's most varied collection of radio-astronomy facilities, including the Very Large Array (VLA) and, in the future, the 25-Meter Millimeter-Wave Radio Telescope.

The three remaining Centers receive federal funding through other channels:

• High Altitude Observatory of the National Center for Atmospheric Research, funded by NSF's Atmospheric Sciences Division for studies of solar physics;
• The recently completed 3-m Infrared Telescope Facility on Mauna Kea, operated as a national facility by NASA; and
• The Space Telescope Science Institute, recently established by NASA at The Johns Hopkins University and charged with the future scientific direction of Space Telescope (ST), together with oversight of support for ST users.

The continued health of U.S. astronomy depends on the proper balance between programs carried out at private and state observatories and those at the National Astronomy Centers. The National Center programs must weigh fairly in this balance for a number of reasons. The Centers furnish the only facilities available to all the nation's astronomers without the necessity of contract or grant support. They are able to provide resources on the scale needed to operate the largest telescopes as nationally accessible facilities. Because the user community is able to exert a decisive influence on the initiation and guidance of scientific research efforts, the National Center programs are highly responsive to the needs of the astronomical community. Finally, the National Centers design and develop instrumentation that later benefits research groups across the nation. It is thus essential to preserve and strengthen the research capabilities of the National Astronomy Centers.

The decade of the 1970's saw a major expansion and upgrading of the observational facilities operated by the National Astronomy Centers for the benefit of the astronomical community. During the 1980's, however, the operations budgets of those centers serving as sites for additional new instruments will have to be increased if the potential of these new facilities is to be realized. Although in some cases economies could be effected through the closing of older Center facilities as new ones become operational, it must also be borne in mind that telescopes do not "wear out." When equipped with modern instrumentation and detectors, older telescopes will continue to provide excellent research opportunities at low cost. Moreover, research opportunities for many Center facilities, both new and old, will expand during the coming decade in response to the scientific challenges to be presented by the next generation of astronomical observations from space. Meeting these challenges will require a vigorous, coordinated program of observations from ground-based observatories covering both hemispheres of the sky. Critically important in such a program will be the contribution of the Cerro Tololo Inter-American Observatory; this facility and the privately operated Las Campanas Observatory are the only two major southern-hemisphere observatories regularly available to U.S. astronomers.

The Committee believes that the National Centers must have strong resident scientific staffs to initiate and guide instrument-development efforts, to assess the performance of observatory equipment, and to assist visiting scientists. To maintain their expertise, such staff members should be encouraged to engage in forefront research of their own and given adequate opportunities to do so.

The National Centers face a difficult task in responding to the diverse needs of a heterogeneous user community. They will continue to need the strong support and encouragement of sponsoring federal agencies in the decade ahead. The Centers must be funded at a level that not only provides for the maintenance of existing facilities and staff but also permits the acquisition of appropriate new equipment in addition to the major capital expenditures recommended by the Astronomy Survey Committee.

I. THE 25-METER MILLIMETER-WAVE RADIO TELESCOPE

The Astronomy Survey Committee supports the construction of a 25-Meter Millimeter-Wave Radio Telescope as provided in the long-range plan of the Astronomy Division of the National Science Foundation and as had been recommended, in an earlier form, in the Greenstein report (*Astronomy and Astrophysics for the 1970's*, National Academy of Sciences, Washington, D.C., 1972). The rapid completion of this facility and its associated instrumentation will be important to continued U.S. progress and leadership in millimeter-wave astronomy during the coming decade.

Studies of star formation, mass loss from stars, the interaction of galaxies with their environment, changes in the nonthermal luminosity of energetic objects, and other important astrophysical processes provoke questions that can be best addressed by observations made at millimeter wavelengths, from 1 to 10 mm, or 30–300 GHz in frequency. For example, the study of star formation benefits from millimeter-wave observations because even optically opaque interstellar clouds are transparent to radiation at millimeter wavelengths, and many of the emission lines of the molecules found in clouds lie in the millimeter spectral region. These molecular radiations provide detailed information about the obscured regions containing the molecules, including the extent, motions, densities, and temperatures of interstellar clouds. This information—particularly when supplemented by the results of infrared photometry and spectroscopy—can provide remarkable insights into the star-formation process itself.

The discovery at millimeter wavelengths of more than 50 molecular species in interstellar clouds has generated a new field of astronomical research, astrochemistry, directed toward understanding the formation and chemistry of interstellar molecules and elucidating the role they play in the collapse of interstellar clouds to form stars.

Millimeter-wave continuum observations provide an important probe of the Universe as a whole. For example, x-ray observations show that many clusters of galaxies are pervaded by diffuse gas having temperatures between 10^7 and 10^8 K; millimeter-wave observations of the microwave background toward such clusters may reveal depressions in the intensity arising from inverse-Compton scattering of the microwave background by the electrons in the hot gas. X-ray and millimeter-wave observations of the same objects can be combined to provide a determination of the value of the Hubble constant that is independent of other methods.

Because of the importance of such studies to astronomy, the Committee believes it essential to undertake a greatly expanded program of millimeter-wave astronomy during the coming decade. Unfortunately, present U.S. single-dish millimeter-wave telescopes have inadequate angular resolution and collecting area, are not sufficiently precise to allow observations at the shortest wavelengths desired, and have restricted sky coverage because of their northerly locations. The 25-Meter Millimeter-Wave Telescope will largely remove such restrictions. The Committee notes that Great Britain, France, West Germany, and Japan are also planning or constructing major new millimeter-wave facilities; these initiatives reflect a worldwide recognition of the importance of this area of research. The United States 25-m telescope will be larger, or will have greater sky coverage, or will operate at shorter wavelengths than these other new instruments, permitting the United States to maintain its leadership in this exciting and highly productive field.

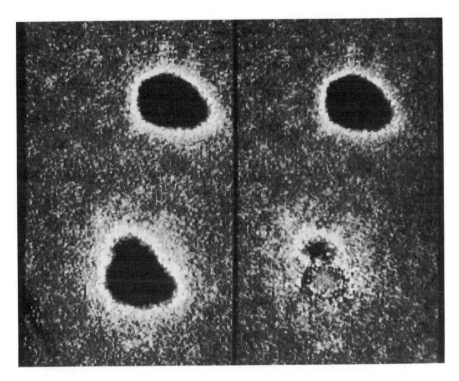

Digitized image-tube images of the two components of a binary QSO, showing (left) the two components and (right) the northern image and the southern image after a scaled version of the former was subtracted from it. The image remaining at the lower right shows the intervening galaxy that is acting as a gravitational lens to produce the binary QSO. (Photo courtesy of A. N. Stockton, University of Hawaii)

5

Prerequisites for New Research Initiatives

In order to be effective, the recommended new research initiatives for the coming decade must be supported by a set of Prerequisites; these are essential for the success of all the major programs but are inexpensive by comparison. Although significant support already exists for each, the Committee strongly recommends substantial augmentations in the following areas, in which the order of listing carries no implication of priority:

A. Instrumentation and detectors,
B. Theory and data analysis,
C. Computational facilities,
D. Laboratory astrophysics, and
E. Technical support at ground-based observatories.

The present chapter discusses the importance of these support activities to the major research directions of the 1980's.

A. INSTRUMENTATION AND DETECTORS

The Astronomy Survey Committee recommends significantly expanded support for instrumentation and detectors during the coming decade as the most cost-effective way to increase the capabilities of both new and existing telescopes.

The past decade has seen spectacular advances in detector and

123

instrumentation technology, resulting in two-dimensional array detectors for optical, ultraviolet, and x-ray wavelengths, and large gains in sensitivity in virtually all wavelength regions. Further development of this technology on all fronts is imperative for the efficient utilization of both ground-based telescopes and space facilities. Improvements in ancillary instrumentation, such as spectrometers, interferometers, and radio-frequency local oscillators, will also be important. As in the past, support for all of these activities should come from a broad base, including the National Sciences Foundation (NSF) grants program, the National Aeronautics and Space Administration's (NASA's) Research and Analysis program, and funding provided through the National Astronomy Centers. The potential gains are enormous, and the cost is small compared with that of major new facilities.

A major detector-development program also exists in the military community, especially for infrared detectors. Rapid declassification of the results of military research and development in this area would be of immense value to astronomy. An efficient means of transferring such declassified technology to the astronomical community is urgently needed.

In many instances a technological breakthrough is only the first step—one must next ensure an adequate supply of new components to the astronomical community at reasonable cost. Commercially available devices may not be well suited to astronomical needs, which usually include maximum sensitivity together with the lowest possible background noise. Often, however, commercial devices can be modified effectively and inexpensively. University laboratories and private industry can contribute to these developments, and such initiatives should be encouraged.

The most critical needs for new detectors and instrumentation during the 1980's are highlighted below; a more complete discussion of these and other projects can be found in the Panel Reports, to be published as a Volume 2 of this survey.

1. Infrared array detectors. Infrared imaging and spectroscopy from the ground and from space would be enormously advanced by the introduction of array detectors in the 1–30-μm range, which are now being developed for military purposes but which have not yet become astronomical tools. They should be reviewed for timely declassification and adaptation to infrared astronomy.

2. Charge-coupled device (CCD) array detectors for the optical, ultraviolet (UV), and x-ray regions. With nearly 100 percent quantum efficiency, high geometric stability, and relatively low readout noise, CCD arrays can be nearly perfect detectors in the optical region; for optical-astronomy applications, however, the devices should have larger area, larger pixels, lower noise, improved blue response, and greater adaptability for large, close-packed mosaics than at present. The wavelength coverage of such detectors can be extended to the UV region by the application of high-efficiency phosphor coatings, although this should be tested further.

Two-dimensional solid-state array detectors are of special value in x-ray observations, since the amount of charge created in the detector is a measure of the energy of each incident photon, so that imaging and low-resolution spectroscopy can be carried out simultaneously. Further development should pursue the goal of detecting each individual photon with an energy resolution approaching the theoretical limit.

3. Millimeter- and submillimeter-wavelength radio receivers. At high radio frequencies (including those sometimes considered to lie in the far-infrared region), the greatest need is for more sensitive coherent detectors. Cryogenic metal-semiconductor junctions (Schottky barriers), InSb mixers, and superconducting junctions will all require further development and increased funding during the 1980's.

4. Gamma-ray detectors. Particular needs include greatly increased flux sensitivity over the entire range, better energy resolution for emission-line spectroscopy, and higher angular resolution to correlate gamma-ray sources more precisely with sources observed at other wavelengths.

5. Cosmic-ray detectors. High sensitivity and resolution are needed to investigate elemental and isotopic compositions throughout the entire periodic table, particularly for heavy elements, whose abundances are extremely low. Devices of very large area are needed to measure the low fluxes expected at energies in the tera-election-volt range.

6. Optical and ultraviolet instrumentation. Greatly increased efficiency in multiobject optical spectroscopy of faint objects can be achieved using multichannel spectrometers. Increased attention should also be given to the use and development of highly reflective optical coatings at all wavelengths, notably in the far UV region.

B. THEORY AND DATA ANALYSIS

The Astronomy Survey Committee recommends support for an expanded program of theoretical astrophysics and data analysis during the coming decade in response to the rich accumulation of data expected from both ground-based and spacecraft observatories.

Astronomy is a field in which new observational discoveries are often unanticipated; their interpretation and eventual understanding require the incorporation of an ever-increasing range of physical processes and concepts into astronomy. Theoretical modeling, usually with the aid of modern computers, is necessary to extrapolate the results of laboratory experiments into totally unfamiliar astrophysical environments. Without extensive and imaginative theoretical analysis, the observational data by themselves are often without apparent pattern or meaning. Furthermore, the new facilities to be put into operation during the 1980's promise to produce data of unparalleled richness and complexity, necessitating ever more sophisticated theoretical interpretation. The Survey Committee thus shares the view of the Panel on Theoretical and Laboratory Astrophysics that the use of these new facilities will stimulate a correspondingly increased level of theoretical activity. Funding agencies should therefore prepare to respond to an increased demand for the support of theoretical astrophysics.

Theoretical astronomical research of the broadest scope has been supported primarily by the NSF Astronomy Division. Such support is critical to astronomy and should be increased. A level of funding 50 percent greater than the present level, as a fraction of NSF astronomy support, is a desirable goal. Implementation could be strengthened by the establishment of a program director for theoretical astrophysics, as recommended by the Panel.

It is inappropriate that the NSF Astronomy Division bear the burden of supporting the basic theory needed for a balanced program in space astronomy. We recommend that NASA establish a strong, broad program in theoretical astrophysics comparable in scope with the NSF program, as NASA has already done in the area of theoretical solar–terrestrial physics.

It is important that this support include a grants program for theory designed to further NASA's overall mission for space-science research but that is funded independently of specific instrumental programs. Such a grants program is necessary to encourage theoretical studies that relate results of different space-science missions to each other and to the results of ground-based astronomy, as well as to incorporate them fully into the conceptual framework of astronomy and physics. At the same time, however, NASA should adequately support the analysis and interpretation of data obtained from specific missions; since this activity naturally includes theoretical interpretation, a significant fraction of NASA support for theory should continue to be funded through channels identified with specific instrumental programs.

The Committee believes that the effectiveness of the National Astronomy Centers would be enhanced by the presence of strong in-house theoretical groups. These groups should support the activities of the Center user communities and should have a number of permanent staff theorists above the critical minimum, as do the Department of Energy national laboratories for research in high-energy physics; it is particularly important that the newly established Space Telescope Science Institute have a strong in-house theoretical staff. Such theoretical groups should be built up carefully in order to ensure appointments of the highest quality.

C. COMPUTATIONAL FACILITIES

The Astronomy Survey Committee recommends that the acquisition of minicomputer and enhanced-minicomputer systems by the U.S. astronomical community be substantially accelerated. About 30 such systems, to be replaced by more advanced systems at intervals of approximately 6 years, are required for data reduction, image processing, and theoretical calculations during the coming decade. However, as some forefront theoretical problems will still require the largest and fastest computers available, the Committee also recommends that federal agencies take steps to provide continued access of the U.S. astronomical community to such machines.

The increasing use of digital imagery in astronomy is already producing a flood of valuable data requiring the extensive use of computers for reduction and analysis. By about 1985, most major telescopes will be equipped with CCD array detectors, the Very Large Array will be in full operation in both continuum and spectral-line modes, Space Telescope will be launched to carry out a rich observational program in optical and UV astronomy, and there will be increased use of microdensitometers to generate digital images from photographic plates. All together, as many as a million images may be generated each year. If the scientific content of these images is to be extracted efficiently, it is imperative that astronomers improve their capabilities for image reduction and analysis. Advances in computer technology now make it possible to exploit the scientific opportunities afforded by new imaging technologies at modest cost.

We stress that image processing is only part of the total picture: the inherent complexity of astrophysical phenomena together with the greatly improved ability to measure these phenomena require increasingly sophisticated analytical techniques for the interpretation of observations. Astronomical phenomena typically involve an intricate interplay of several strongly nonlinear effects. In such situations, theoretical modeling by digital computation often provides the only practical approach to understanding. The demand for computational facilities for theoretical modeling will increase in parallel with the demand for image-processing facilities in the 1980's. Here again, advances in computer technology allow the important scientific opportunities for detailed theoretical modeling to be exploited at a modest cost.

Computational capability will continue to be provided through three main sources: university computer centers, large computers operated by national laboratories and observatories, and dedicated minicomputers and superminicomputers. Computers at national laboratories and observatories constitute a unique and vital resource for handling some theoretical problems, and their shared use by astronomers should be encouraged and expanded. In recent years, however, new computer technology has brought about a dramatic shift in the capability and cost-effectiveness of minicomputers in comparison with the large central computers operated by university computer centers. Minicomputers and superminicomputers are now able to handle almost all image processing, as well as many theoretical calculations. Their cost-effectiveness compares favorably with that of the computers operated by national laboratories and observatories and often exceeds by a large margin that of many university computer centers. Furthermore, they provide the flexibility and interactive capability necessary for creative interpretation of observational data and theoretical results. (These new computational opportunities have been recognized not only in the United States but also abroad; in 1979, for example, the United Kingdom's Science and Engineering Research Council established a network of six linked minicomputers to meet the image-processing requirements of U.K. astronomers during the 1980's.) In order for these facilities to operate effectively, attention and support must also be given to the standardization and sharing of software for the most commonly encountered calculations.

The Committee believes that the acquisition by the U.S. astronomical community of dedicated minicomputer- and supermini-computer-based computational systems should be substantially accelerated. However, the Committee also recognizes the existence of computational problems of outstanding importance to astrophysics that can be attacked only with the most powerful computers. We urge NSF and NASA to help ensure that qualified astrophysicists continue to have access to the largest and most sophisticated computing equipment available.

D. LABORATORY ASTROPHYSICS

The Astronomy Survey Committee recommends expanded support for the laboratory measurements of atomic, molecular, and nuclear properties needed for the interpretation of nearly all astronomical observations. Federal agencies should furthermore coordinate their efforts in providing such support and should take explicit account of the need for laboratory astrophysical data in the planning of future scientific activities and missions.

The accuracy with which physical conditions can be inferred from spectroscopic observations depends directly on the breadth and precision of the data available for atomic, molecular, and nuclear processes. The laboratory research that yields these data has not been funded adequately during the past decade, with increasingly damaging consequences for astronomical investigations. For example, uncertainties in solar opacities have hindered precise predictions of solar neutrino emission, thereby throwing doubt on an important scientific test of our understanding of stellar structure and evolution; lack of accurate molecular data is an obstacle to the quantitative understanding of physical and chemical processes in interstellar clouds and prevents confident use of remotely sensed spectra for an understanding of planetary atmospheres.

The funding of laboratory astrophysics has suffered from an erratic pattern of support for atomic and molecular physics generally. This trend has resulted in a severe decline in the number of atomic and molecular physicists responding to the growing needs of astronomy. A strong stimulus is needed to attract a greater fraction of existing laboratory talent to problems related to astronomy.

The 1980's will see the deployment of powerful new instruments for astronomical spectroscopy and a resulting unprecedented growth in the quality and variety of astronomical spectroscopic data. An allied growth in laboratory astrophysics will be required to utilize these new data fully. Expanded support by NSF will be needed in the years ahead. We also recommend that NASA make an increased effort in its mission-planning and Research and Analysis programs to fund the laboratory measurements in basic physics and chemistry that are needed for the interpretation of astronomical observations from space. Of particular importance to NASA missions are studies of highly ionized atoms relevant to UV and x-ray spectroscopy, of molecular physics relevant to infrared and radio spectroscopy, and of heavy-nuclei interaction cross sections relevant to cosmic-ray isotope and element spectroscopy.

The Committee believes that NSF, NASA, and the Department of Energy should develop coordinated programs for the support of research in atomic, molecular, and nuclear physics, which are of interest to astronomy and to other branches of science, and should also support interdisciplinary workshops and symposia. We furthermore urge NSF to increase its efforts to encourage laboratory astrophysics by coordinating the relevant activities in the Astronomy and Physics Divisions.

Finally, we urge the National Bureau of Standards (NBS) to increase its support of basic studies in atomic and molecular physics and of laboratory astrophysics. These NBS programs have contributed importantly to many fields of pure and applied physics as well as to astronomy, and they provide an environment in which vital cross-fertilization among these fields can occur.

E. TECHNICAL SUPPORT AT GROUND-BASED OBSERVATORIES

The Astronomy Survey Committee recommends expanded support for the technical personnel needed to ensure the development, maintenance, and enhancement of modern astronomical instrumentation at ground-based observatories.

The efficient progress of optical, infrared, and radio astronomy in the coming decade is dependent on improvements in the instrumental capability and continued productivity of ground-based observatories. Recent advances in instrumental techniques and control systems, together with those foreseeable in the near future, represent a potential for greatly increasing the capability of both large and moderate-sized telescopes. These advances, however, are worthless to astronomy unless observatories have the means to develop, implement, and maintain the working instruments that take advantage of them.

The present shortage of technical personnel who are able to assist in the development of instrumentation, as well as in its maintainance and improvement, is a major difficulty. We urge that funds be allocated to support such personnel, with the strong proviso that the award of such funds should proceed through the normal peer-review channels and that the basis of the award be the excellence of the science proposed with the instrumentation in question.

Small institutions and small telescopes should not be excluded from such awards but should be supported if the scientific proposals have merit. The modernization of small instruments at reasonably good sites can help alleviate the pressure on large telescopes in an extremely cost-effective way, and it provides a unique resource for small-scale but highly innovative research.

The Multiple-Mirror Telescope of the Smithsonian Institution and the University of Arizona. (Photo courtesy of G. McLaughlin, Lunar and Planetary Laboratory)

6

New Programs

The New Programs recommended by the Committee for approval and funding during the coming decade have been divided into three categories according to the scale of resources required for their completion.

A. *Major New Programs* The Committee believes that four major programs are critically important for the rapid and effective progress of astronomical research in the 1980's and is unanimous in recommending the following order of priority:

1. An *Advanced X-Ray Astrophysics Facility* (AXAF) operated as a permanent national observatory in space;
2. A *Very-Long-Baseline* (VLB) *Array* of radio telescopes designed to produce images with an angular resolution of 0.3 milliarcsecond;
3. A *New Technology Telescope* (NTT) of the 15-m class operating from the ground at wavelengths of 0.3 to 20 μm, with relevant design studies to be undertaken immediately; and
4. A *Large Deployable Reflector* (LDR) in space, for spectroscopic and imaging observations in the far-infrared and submillimeter regions of the spectrum that are inaccessible to study from the ground.

B. *Moderate New Programs* In rough order of priority, these are:

133

1. An augmentation to the National Aeronautics and Space Administration (NASA) Explorer program,
2. A far-ultraviolet spectrograph in space,
3. A space VLB interferometry antenna in low-Earth orbit,
4. The construction of optical/infrared telescopes in the 2–5-m class,
5. An Advanced Solar Observatory in space,
6. A series of cosmic-ray experiments in space, and
7. An astronomical Search for Extraterrestrial Intelligence (SETI).

C. *Small New Programs* The program of highest priority is:
- A 10-m submillimeter-wave antenna.

Other programs of outstanding scientific merit, in which the order of listing carries no implication of priority, are as follows:
- A spatial interferometer for the mid-infrared region,
- A program of high-precision optical astrometry, and
- A temporary program to maintain scientific expertise at U.S. universities during the 1980's through a series of competitive awards to young astronomers.

A. MAJOR NEW PROGRAMS

1. Advanced X-Ray Astrophysics Facility

The Astronomy Survey Committee recommends the construction of an Advanced X-Ray Astrophysics Facility (AXAF) to be operated as a permanent, national observatory in space and urges the National Aeronautics and Space Administration (NASA) to begin its development in time to ensure AXAF operation by the end of the decade.

In less than 20 years, x-ray astronomy has advanced from the discovery of the first extrasolar x-ray source (through a brief, exploratory rocket experiment) to the detailed study of thousands of Galactic and extragalactic sources with the image-forming x-ray telescope on the *Einstein* (HEAO-2) Observatory. X-ray observations have revealed important new classes of astronomical objects and have dramatically advanced our understanding of astrophysical processes in virtually every field of astronomical research, from stellar physics to studies of quasars and the formation and evolution of galaxies. Finding x-ray observations to be of vital significance in their work, many astronomers from other fields participated in planning and interpreting the *Einstein* observations during its two years of operation, and x-ray observations have now attained an importance in contemporary astronomy comparable with those in other wavelength regions. An urgent scientific need therefore exists for a long-lived satellite observatory with capabilities for x-ray astronomy that complement those of Space Telescope (ST) in the optical/ultraviolet region and those of the Very Large Array in the radio region of the spectrum.

The AXAF will fulfill that need with an instrument that utilizes the same basic principles that were tested and proved in the *Einstein* mission but which is capable of providing up to a hundredfold greater sensitivity for the study of faint stellar or quasistellar objects and a tenfold increase in angular resolution for the study of structure in extended objects. Major improvements will be achieved in spectroscopic sensitivity and resolution, and a capability for sensitive polarimetry will also be provided. The Space Shuttle will provide the means for launching AXAF, maintaining it in orbit, and retrieving it for major refurbishments. Thus, like ST, AXAF will be a national facility that can meet fundamental needs of astronomy for a decade or more.

AXAF will permit the observation of sources with x-ray luminosities as small as 1 percent of the Sun's total luminosity lying in the farthest reaches of our Galaxy, as well as the study of all the individual high-luminosity x-ray sources in the hundreds of galaxies of the Virgo cluster. The composition and dynamics of extended sources such as supernova remnants, galaxy halos, and clusters of galaxies can be revealed by spectroscopic and polarimetric observations of high angular resolution. The great sensitivity of AXAF will permit investigation of x-ray galaxies and clusters of galaxies out to distances so large that the effects of evolution in the early Universe should be apparent. Because of its power and versatility, AXAF will profoundly influence and enhance the development of nearly all areas of Galactic and extragalactic astronomy.

The Committee also suggests that NASA consider the establishment of special institutional arrangements similar to those embodied in the Space Telescope Science Institute, to provide scientific guidance for the development and maintenance of AXAF, to manage the scientific direction of the mission during orbital operations, and to facilitate the participation of the scientific community in the acquisition and interpretation of x-ray observations. Consideration should be given, as in the case of the Space Telescope Science Institute, to appropriate international participation.

2. A Very-Long-Baseline (VLB) Array of Radio Telescopes

The Astronomy Survey Committee recommends the construction of a ground-based Very-Long-Baseline (VLB) Array of radio telescopes designed to produce images with an angular resolution of 0.3 milliarcsecond. Because the Array utilizes proven technology, this project may be begun immediately after completion of final management and design studies.

Extraordinarily high angular resolution is now possible at radio frequencies. Precision atomic clocks, more sensitive and reliable receivers, high-speed tape recorders, sophisticated image-processing techniques, and modern antennas now make it feasible to build a radio array with the angular resolution of a telescope covering an entire continent. This may be done by synchronizing the operation of about ten widely spaced antennas of approximately 25-m diameter, whose outputs are recorded and later combined in a central computer.

This VLB Array will produce high-quality radio images capable of resolving features down to 0.3 milliarcsecond (the size of a dime in New York as seen from Los Angeles). This is a hundred times better angular resolution than that of any other image-forming telescope at any wavelength and will yield detailed new radio images of a wide range of astronomical objects at the frontiers of modern astrophysical research. These include quasars and the nuclei of galaxies, features of interstellar molecular clouds, the center of our Galaxy, and a variety of energetic Galactic objects such as x-ray, binary, and flare stars. The high angular resolution of the VLB Array will permit the direct study of small-scale structure surrounding the central regions of quasars and stars in the process of formation. Through the method of statistical parallaxes, it will furthermore permit direct measurements of distances to many objects throughout our Galaxy and even to some in nearby galaxies. The VLB Array can also be applied to important problems in Earth science (including precision geodesy and geophysics), to the navigation of interplanetary spacecraft, and to tests of the General Theory of Relativity.

Although the VLB Array is a complex and sophisticated instrument, it will make use of proven concepts and instrumentation. Construction should begin immediately upon completion of management and design studies with the building of the antennas and the development of the data-reduction system and other instrumentation. Collaboration with groups in other countries, particularly in Europe and North America, would improve the performance of the instrument by increasing the resolution even further (particularly in the north–south direction) and by improving the image quality at low declinations.

3. A New Technology Telescope (NTT) of the 15-Meter Class

The Astronomy Survey Committee recommends the construction of a New Technology Telescope (NTT) of the 15-m class on the ground for observations in the optical and in the near- and mid-infrared regions of the spectrum (0.3- to 20-μm wavelength). *The design studies needed before the NTT can be constructed are of the highest priority and should be undertaken immediately.*

Recent progress in optical fabrication techniques, design concepts, and electronics now make it possible to build a large optical/infrared telescope at a cost much lower than was possible a decade ago. Such a New Technology Telescope (NTT), having a diameter of approximately 15 m, will increase our observing capabilities in the critical 0.3–20-μm spectral region in two important ways. First, throughout this spectral region, the vast area of the mirror—nine times larger than that of the 5-m Mt. Palomar reflector—will collect light at a rate exceeding the combined capabilities of the world's 20 largest existing optical telescopes, furnishing the image brightness needed for a new generation of spectroscopic observations. Second, at a good site, the NTT should frequently achieve 0.3-arcsec resolution at 20-μm wavelength; in the absence of special speckle or interferometric techniques, this angular resolution can generally be surpassed only by Space Telescope, at much shorter wavelengths. For many infrared applications, NTT's combination of large collecting area and high angular resolution will lead to a tenfold increase in limiting sensitivity and a hundredfold increase in speed over present capabilities.

NTT's large collecting area will make it an enormously powerful tool for the spectroscopy of faint astronomical sources. For example, detailed spectra of faint, old stars on the fringes of our Galaxy will outline for us the early history of element building and nucleosynthesis during the birth of our Galaxy. Similar observations in nearby dwarf spheroidal galaxies, satellite systems of the Milky Way, will tell us how the formation and early evolution of these small sister galaxies differed from that of our own Galaxy. In yet a third such study, astronomers will determine the compositions and motions of the equally old swarms of globular clusters that surround many neighboring galaxies. These clusters, currently believed to be products of the initial galactic collapse phase, hold still further clues to the mysteries of galactic birth and evolution. All of these are threshold problems, in the sense that there are no brighter objects nearby that are suitable for study; without the light-gathering power of NTT, an insufficient number of photons can be collected to mount the decisive spectroscopic programs needed to address these important issues.

In addition, spectroscopic observations of the most distant galaxies and quasars will greatly advance our understanding of cosmic evolution. For example, NTT studies of quasar absorption lines will permit measurements of the distribution and composition of intergalactic gas as it existed very early in the history of the Universe. These measurements offer the exciting possibility of tracing back the origin of the chemical elements and the birth of clusters and superclusters of galaxies to a time much earlier than we can currently see directly; indeed, we should be able to study the large-scale properties of the Universe when it was only one quarter as old as it is now. Such studies are completely impossible with present optical telescopes because of their inadequate collecting areas; the faintness of quasars, in particular, requires the enormous light-gathering power of NTT for systematic spectroscopic study.

The combination of NTT's high spatial resolution with sufficient photon-collecting power to achieve very high spectral resolution will also permit definitive studies of molecular clouds and obscured protostars in the near- and mid-infrared regions of the spectrum. The fundamental vibration–rotation transitions of molecules are found primarily in the 2–10-μm wavelength region. The study of these and of molecular rotational transitions and continuum radiation will lead to a much better understanding of the composition, abundances, excitation, and dynamics of collapsing gas clouds. NTT will provide a probe of gas dynamics in regions of star formation by permitting the examination of optical and infrared spectral lines at very high spectral resolution on an exceedingly fine spatial scale. Present observations indicate that the brightest protostellar candidates are about 1 arcsec in diameter in the 2–10-μm wavelength region, whereas the diffraction limit of a filled-aperture 15-m telescope is about 0.03–0.15 arcsec over the same range; thus, the use of NTT with interferometric techniques will permit detailed study of the geometry and structure of such objects. In addition, NTT's high spatial resolution will allow the isolation and study of individual source components. All of these studies will be complementary to those that can be carried out by the Shuttle Infrared Telescope Facility (SIRTF), which has high sensitivity but low angular resolution; by ST, which has high angular resolution but limited collecting area; and by the Large Deployable Reflector in space, which will be designed for far-infrared work at wavelengths longer than 20 μm.

Because most of the known objects in the Universe either emit visible or infrared radiation or are associated with objects that do, optical and infrared spectroscopy provide powerful and versatile techniques for investigations of the Universe. As the world's most capable instrument for such spectroscopic observations, NTT will be an extraordinarily productive facility. The Survey Committee finds the scientific importance of NTT to be equal to that of any other facility considered and regards it as one of the cornerstones of the recommended research program for the 1980's.

4. A Large Deployable Reflector in Space

The Astronomy Survey Committee recommends the construction of a Large Deployable Reflector (LDR) of the 10-m class in space to carry out observations in the far-infrared and submillimeter regions of the spectrum that are inaccessible from the ground. Design studies for such a facility should begin at once.

Much of the matter in the Universe is relatively cool, from cool stars at a few thousand degrees to dense interstellar clouds at tens of degrees. Radiation from these objects lies in the infrared and submillimeter regions of the spectrum, at wavelengths from a few to a few hundred micrometers. Instruments in space are essential for observations at wavelengths between the atmospheric windows and at wavelengths beyond 20 μm, for which the elimination of atmospheric absorption and background thermal noise is critical. SIRTF will offer a thousandfold improvement in our ability to detect infrared sources, bringing many millions of them into view. However, the requirements of cryogenic cooling put a practical limit on the aperture of the SIRTF telescope; as a result, it cannot collect enough photons for high-resolution spectroscopy, and its angular resolution is high only at shorter infrared wavelengths.

A LDR in space of approximately 10-m diameter, for observations at the longer infrared and submillimeter wavelengths inaccessible from the ground, is needed to collect enough photons for high-resolution spectroscopy and to provide high angular resolution at these long wavelengths. Such a telescope could carry out detailed morphological and spectroscopic studies of all the far-infrared sources discovered in the forthcoming Infrared Astronomy Satellite (IRAS) all-sky survey and of the brighter objects discovered by SIRTF. Because mirror-figure and pointing requirements are a hundred times more relaxed than for ST, a 10-m-class LDR capable of arcsecond resolution at 20-μm wavelength should be less expensive.

A number of important scientific problems are uniquely accessible to such a LDR in space. For distances less than 500 parsecs, the projected beam diameter will be less than 1000 astronomical units. Direct measurements of the sizes of nearby clouds collapsing to become stars will thus be possible at far-infrared wavelengths, which can penetrate the surrounding clouds of dust that invariably obscure small-scale features at optical wavelengths. In addition, the wavelength regions accessible to a LDR contain spectral lines of atoms, ions, and molecules that reflect a wide range of astrophysical conditions. Studies of these features will yield otherwise unobtainable information about the structure and dynamics of planetary atmospheres; the heating, cooling, and chemical composition of the interstellar medium; and—because of the penetrating power of long-wavelength radiation—chemical abundances in the highly luminous but optically obscured nuclei of active galaxies.

The sensitivity and high angular resolution of a LDR will also make it possible to study newly forming stars in optically obscured regions of nearby external galaxies, enhancing our understanding of galactic evolution and of the dynamical processes that stimulate star formation. Such an instrument can also probe the structure of the early Universe and the mechanisms of galaxy formation through studies of small-scale spatial fluctuations in the cosmic microwave background radiation.

The capabilities of a LDR in space will complement those of ST, which is optimized for observations in the ultraviolet, optical, and near-infrared spectral regions, and those of the NTT, which will offer large collecting area and high angular resolution but will be restricted to the atmospheric windows between 0.3- and 20-μm wavelength. The Committee believes that design studies for a LDR should begin at once.

B. MODERATE NEW PROGRAMS

1. Explorer Program Augmentation

The Astronomy Survey Committee recommends an immediate and substantial augmentation to the NASA Explorer satellite program, with the aim of restoring it to at least the healthy real level of effort of 1970.

NASA is now proceeding with four Explorer missions in the areas of astronomy and astrophysics. The Infrared Astronomy Satellite (IRAS) is at present under development; the Cosmic Background Explorer (COBE), Extreme Ultraviolet Explorer (EUVE), and X-Ray Timing Explorer (XTE) are in various stages of final planning in preparation for development. Unless the present Explorer funding level is increased, however, these are likely to be the only Explorer satellites dedicated to astronomical observations that can be flown in the 1980's. The Astronomy Survey Committee believes that such a limitation would present a serious obstacle to the progress of space astronomy during the coming decade.

Today's Explorer budget, as currently charged for mission costs, provides only about half the support in real terms that was available to the Explorer program a decade ago. As a consequence, the flight of new Explorer missions has in recent years fallen much below the rate needed for healthy advance. The rate will decline even more drastically during the early 1980's if present budget levels are not increased. The Astronomy Survey Committee thus recommends an immediate and substantial augmentation to the Explorer program to restore it to at least the real level of effort of 1970.

As emphasized in Chapter 4, the Explorer program has been a vital component of the NASA space-science program for over 15 years, and it promises to continue to provide the best means for pursuing a wide range of scientific problems in the years ahead. Determination of priorities among the most promising individual Explorer mission possibilities in astronomy and astrophysics for the 1980's remains the responsibility of other advisory groups, particularly the Space Science Board's Committee on Space Astronomy and Astrophysics (CSAA). However, among the scientific areas that at present appear to offer special promise for additional Explorer-class missions are the following, in which the order of listing carries no implication of priority:

• A spectroscopic study of physical conditions and element abundances in a wide variety of x-ray sources. Such a study could address one or more of the regimes of spectroscopy that may not or will not be addressed by AXAF, e.g., observations of spectral lines (such as Fe lines) with exceptionally high spectral resolution or with the spatial resolution permitted by large apertures, the study of spectral lines emitted by newly synthesized matter and of cyclotron-resonance features at energies above 10 keV, and wide-field spectroscopic studies of the interstellar medium. These investigations would yield important new information on the structure and composition of young supernova remnants and the interstellar medium; the structure and dynamical behavior of the coronas of nearby stars; the nature of the plasmas associated with compact x-ray sources in Galactic x-ray binaries and in quasars; and the distribution, composition, and origin of the hot intergalactic gas pervading clusters of galaxies.

• A study of the isotopic and elemental composition of low-energy Galactic cosmic rays and solar energetic particles in the interplanetary medium. A primary aim of such an investigation is an accurate determination of the isotopic composition of the elements through nickel in a direct sample of contemporary solar and Galactic interstellar matter. In addition, it should be possible to analyze the composition of solar energetic particles through uranium and to study the processes that accelerate particles on the Sun and in the interstellar medium. These measurements hold the key to understanding the processes that synthesized and accelerated both solar and Galactic matter.

• A study in soft x rays (preferably with moderate-resolution spectroscopic capability) of those objects now known to radiate predominantly in the 100–2000-eV energy range, including cataclysmic variable stars, AM Herculis-type systems, RS CVn binary stars, stellar coronas, isolated white dwarfs, central stars in planetary nebulae, hot neutron-star remnants of recent supernovae, and x-ray pulsars. Such an investigation should include extended searches for, and studies of, regular and quasi-regular pulsations as well as aperiodic variability, studies of binary orbital light curves, measurements of spectral-line ratios (to determine temperatures and densities in the hot, emitting plasmas), and observations correlated with measurements at other wavelengths. The results of these studies would have an important bearing on our understanding of the evolution of compact binary systems, the composition and cooling of isolated very hot stars, and stellar activity and variability cycles.

• A study of high-energy transient phenomena, particularly observations of cosmic and solar gamma-ray bursts up to 10 MeV with high resolution and sensitivity, in conjunction with a program of wide-field x-ray imaging with fine angular resolution. Such a combination will provide both accurate timing and location of the burst sources from the same spacecraft and very probably lead to their identification. The cosmic gamma-ray observations will also provide probes of nuclear and electromagnetic processes in compact objects and measure spectral features that have been reported in burst measurements, including cyclotron-resonance features, red-shifted 511-keV photons, and nuclear lines. Continuous observations of solar gamma-ray lines may also be possible from the same spacecraft; such observations, carried out over a several-year period bracketing a solar maximum and with the highest energy resolution obtainable, would greatly advance our understanding of solar flares.

• A study of the physical processes that deposit both energy and momentum into the solar corona and the solar wind, through measurements of the structure, expansion velocity, electron-density distribution, transition-region plasma density, and magnetic properties of the corona. Such investigations will furnish essential new information on the detailed relation of coronal structure to x-ray bright points and coronal holes through studies of the fate of newly emerging magnetic-flux tubes; the interaction of coronal plasma with the solar magnetic field in general; and mechanisms for the acceleration of the solar wind, which are best studied in conjunction with simultaneous observations of the evolution of coronal holes, bright points, transients, active regions, and magnetic structures on all scales.

• A study of the interior dynamics of the Sun, as one important component of a more general program to understand the fundamental mechanisms responsible for the solar cycle. The goals of such an interior-dynamics study include measurements of photospheric velocities, as a probe of the solar convection zone and the variation of rotation rate with depth; of radiation from the large-scale convective pattern in the photosphere, to permit correlations with magnetic measurements and inferences concerning the influence of magnetic fields on solar radiative output; of radiation from the entire disk of the Sun over periods ranging from days to the length of a solar cycle, to determine the degree of variations in the so-called "solar constant"; and of magnetic activity in the upper solar atmosphere, to determine the effects of such activity on the atmospheres of both the Sun and the Earth. These results are also expected to play a role in the more general effort to understand the nature of stellar convection and of energy and magnetic-field maintenance.

Two further possibilities merit detailed study by NASA and by other advisory groups for inclusion in the Explorer program. The first is an Explorer satellite to map the Milky Way at moderate angular resolution, both in the lines of selected submillimeter-wavelength transitions thought to be important for the heating and cooling of interstellar gas clouds and in wavelength bands relevant to determinations of temperature and density distributions in cold clouds. The second is the Explorer flight of optical and infrared interferometers capable of submilliarcsecond angular resolution, which would represent an important step toward realization of the long-range program of interferometry recommended for study and development in Chapter 7.

The Committee also notes that several other missions (such as the far-ultraviolet spectrograph in space discussed below) might be carried out within the Explorer program if substantial funding contributions from foreign or other national space programs become available. These and other possible collaborations within the Explorer program should be investigated.

2. Far-Ultraviolet Spectrograph in Space

The Astronomy Survey Committee recommends the launch of a far-ultraviolet spectrograph in space incorporating a 1-m class telescope to be used primarily for high-resolution spectroscopy in the 912–1200-Å spectral region.

In the spectral region between the 912-Å absorption edge of neutral hydrogen and the 1200-Å onset of reflectivity of the ST mirror coatings lie many important features critical to the understanding of the interstellar gas, extended stellar atmospheres, supernova remnants, galactic nuclei, and gaseous halos of planets. Atomic and molecular lines not detectable in the ST spectral region include the resonance lines of O^{+5}, seen in stellar coronas and in interstellar space; O^0 and N^+ resonance lines needed for the determination of nitrogen and oxygen abundances in ionized H II regions; and the lines of H I, H_2, and HD, which are important for understanding the chemistry of molecular clouds and for determining the cosmic D/H ratio. For many other important species (e.g., ^{12}CO, ^{13}CO, N I, Fe II), complementary studies with the proposed instrument and the high-resolution spectrograph on ST are required, neither being sufficient alone. More generally, the value of data returned in the 912–1200-Å spectral region would be enhanced by complementary, concurrent studies both of ionic resonance lines in the 1200–2000-Å region and of spectral features characteristic of hot interstellar plasma extending down to 300-Å wavelength or even less. If suitably designed, a far-ultraviolet spectrograph could also have applications to the study of upper-atmospheric processes in planets and of the hot, ionized torus of gas surrounding Jupiter.

The technology for building a powerful far-ultraviolet spectroscopic facility with a spectral resolution of at least 3×10^4 now exists, but further improvements in the sensitivity of windowless detectors are possible and should be pursued. Interesting scientific goals that should set the cost and scope of the project include: (1) the detection at a resolution of 3×10^4 of a 12th-magnitude unreddened B0 star in a few hours with a signal-to-noise ratio of 20 to 1, to allow detailed studies of Galactic gas up to 8 kpc from the Sun at all Galactic latitudes and in a few lines of sight to the Large and Small Magellanic Clouds; (2) similar detections with longer integration of 14th-magnitude objects for studies of a number of Seyfert galaxy nuclei, as well as of our Galactic halo and halos of other galaxies; (3) the detection of 17th-magnitude objects with lower resolution to include many Seyfert galaxies and several quasars (quasar observations can also provide probes of gaseous halos around a few foreground galaxies); and (4) study of the stability of the ionized gas torus in the Jovian magnetosphere.

As there is considerable interest in a far-ultraviolet spectrograph on the part of European astronomers, NASA should explore the possibility that this project could be partially funded by foreign astronomical groups.

3. A Space VLB Interferometry Antenna

The Astronomy Survey Committee recommends the placement of a space VLB interferometry radio antenna in low-Earth orbit to extend the powerful VLBI technique into space in parallel with the rapid completion of a ground-based VLB Array.

A space antenna in low-Earth orbit will complement and extend the capabilities of the ground-based VLB Array in at least four important ways. First, through provision of a variety of additional baselines, a space antenna will permit much more complete coverage of the Fourier-transform plane, which is the source of all VLBI information on the detailed structure of radio sources. This additional coverage will allow a more complete (and in many cases unambiguous) mapping to be made of many complex sources. As the expansion velocities of radio components in quasars may appear to exceed that of light solely because of incomplete coverage in the Fourier-transform plane, such coverage will in many cases help to test the reality of this effect. More complete coverage in the Fourier-transform plane also produces a cleaner beam, permitting the mapping of sources with low surface brightness. In radio galaxies and quasars, for example, features nearer the compact nucleus tend to emit more weakly than the larger, more developed structures found further away; the capability to study structures with low surface brightness thus permits a more complete study of dynamical processes presumed to be operating at various distances from the central energy sources in these objects.

Second, a space VLBI antenna greatly facilitates the study of radio sources at low declination. Any ground-based VLB Array must rely on the rotation of the Earth to vary the position and length of the baseline employed; for low-declination sources, the resulting baselines lie largely or solely in the east–west direction, thus diminishing greatly the north–south resolution and degrading the quality of the radio map. A space antenna can provide north–south resolution of such sources, allowing full two-dimensional mapping of radio sources at all declinations.

Third, an Earth-orbiting space antenna will permit the study of time variation on much shorter time scales than is possible with a purely ground-based array, which requires up to 24 h to fill in all the baselines necessary to complete a radio map. An antenna in low-Earth orbit would allow mapping of sources in a time that is half of the period of revolution of the satellite (about 1 h) and, thus, the study of source variations on this much shorter time scale. Such a capability would be immediately useful, for example, in studying dramatic relativistic-jet phenomena in SS 433. Many other time-varying sources will also be accessible.

Finally, a space antenna will—in concert with the VLB Array's southernmost stations, southern hemisphere NASA stations, and antennas in other countries—permit the VLBI mapping of the rich southern hemisphere of the sky.

The Committee notes that even the first space VLBI antenna in low-Earth orbit will additionally provide angular resolution that is nearly an order of magnitude greater in solid angle than that of the ground-based VLB Array alone. Moreover, a space VLBI antenna in low-Earth orbit will provide the first step toward the achievement of baselines far longer than can ever be achieved on the Earth itself. This requires the placement of antennas in highly elliptical Earth orbits; in Chapter 7, the Committee recommends the study and development of a more extensive space VLBI system as part of a program of advanced spatial interferometry in the radio, optical, and infrared regions of the spectrum.

4. Construction of Optical/Infrared Telescopes in the 2–5-Meter Class

The Astronomy Survey Committee recommends the construction of optical/infrared telescopes in the 2–5-m class during the coming decade as the ground-based facilities of choice for an extensive range of important observations. The Committee particularly encourages federal assistance for those projects that will also receive significant nonfederal funding for construction and operation.

Optical/infrared telescopes in the 2–5-m class have made critically important contributions to most of our recently acquired knowledge in a number of key areas, including the following:

- Evidence of galaxy evolution from studies of distant galaxies;
- The crucial quasar observations that established the existence of the first known gravitational lens;
- A mass estimate for the dense component of an x-ray binary star, establishing it as the leading black-hole candidate;
- The discovery of quasars in clusters of galaxies, demonstrating that at least some quasars are at the great cosmological distances implied by their large red shifts;
- Dynamical studies of hundreds of galaxies in clusters, showing that the bulk of the matter in the Universe is nonluminous at optical wavelengths and has therefore escaped direct detection;
- Observations providing strong evidence that interstellar shock waves play a role in star formation;
- Demonstration that the old stars in our Galaxy are anomalously rich in oxygen, supporting the hypothesis that the protogalactic gas was enriched by an early generation of massive, short-lived stars;
- Detection of activity cycles in solar-type stars and observations of the modulation of chromospheric features by stellar rotation;
- Discovery of a binary-star system (SS 433) emitting huge streams of matter at about one fourth of the speed of light; and
- Optical confirmation of the existence of pulsars.

In addition, telescopes in the 2–5-m class have furnished essential follow-on observations and identifications of a multitude of objects discovered in other wavelength regions by spacecraft or complementary ground-based facilities. Such telescopes are essential for timely observations of transient phenomena, long-term survey and surveillance programs, general support of space astronomy, and the development of astronomical instrumentation under realistic observing conditions.

All of these scientific tasks and opportunities will assume even greater importance during the coming decade. Recent breakthroughs in telescope technology have radically reduced the cost of construction of optical/infrared telescopes in the 2–5-m class, enabling a wider group of institutions to consider their acquisition than ever before. Such telescopes, equipped with modern instrumentation and detectors, will constitute powerful research tools for the 1980's. The Committee believes that federal funds for such facilities should be awarded on the basis of peer review of scientific merit. Federal agencies should be receptive to proposals from all parties that could make effective use of these telescopes, including private, state, and national institutions.

The Committee particularly encourages the award of federal funds for such telescopes to be located at private and state observatories that can contribute significant nonfederal funding for construction and operation. It applauds the initiative of astronomy groups that are currently seeking private and state funds for this purpose and welcomes private and state agency support of astronomical research. Some portion of the observing time at these facilities should be allocated to outside users, the fraction to depend in general on the level of federal funding. Visitor access is especially important for the larger telescopes constructed in this class.

5. Advanced Solar Observatory in Space

The Astronomy Survey Committee recommends the establishment in space of an Advanced Solar Observatory (ASO)—to be assembled near the end of the coming decade from facility-class instruments developed earlier through the Spacelab program—for simultaneous observations of a number of important solar properties at optical, extreme ultraviolet, x-ray, and gamma-ray wavelengths.

The Solar Shuttle Facility endorsed earlier (Chapter 4, Section D) can achieve a number of the major scientific objectives of space solar physics. However, other important problems—such as the structure of the convection zone, large-scale circulation patterns, transient high-energy phenomena, and long-term evolution of the corona—require observations with a comprehensive set of high-resolution instruments over much longer periods of time than can be obtained in Shuttle flights. The Committee therefore recommends that the Solar Shuttle Facility evolve into a major free-flying observatory, the ASO.

The ASO is currently planned to consist of five major components: (1) the Solar Optical Telescope (SOT) truss containing the SOT itself; (2) a Solar Soft X-Ray Telescope Facility (SSXTF), an EUV Telescope Facility (EUVTF), and white-light and resonance-line coronagraphs to be mounted within the SOT truss; (3) a Grazing Incidence Solar Telescope (GRIST) currently under consideration for development by the

European Space Agency; (4) a Pinhole/Occulter Facility for hard x-ray observations and high-resolution studies of the corona; and (5) a Solar Gamma Ray Telescope. It is hoped that an ASO nucleus (consisting for example of the SOT/SSXTF/EUVTF assembly and GRIST) can become operational in the late 1980's, perhaps assembled upon a space platform. The full ASO, including the Pinhole/Occulter Facility and the Gamma Ray Telescope, will be necessary for the study of transient phenomena during the next solar maximum in the early 1990's.

ASO will be the first solar facility with the design goal of achieving an angular resolution much less than 1 arcsec from the infrared through the x-ray region (0.1 arcsec or better at many wavelengths), together with the spectral resolution needed to detect motions of solar material over the entire span of wavelengths required for study of such material at a wide range of temperatures. The ASO will be able to address nearly all of the problems of contemporary solar physics, including the structure of the solar core; the mechanisms responsible for the solar magnetic and activity cycles; the energy and mass-transport mechanisms operating over the full range of temperatures present in the atmosphere; the basic plasma processes responsible for metastable energy storage, magnetic reconnection, and particle acceleration in solar flares and related nonthermal phenomena; and the processes involved in the heating and acceleration of the solar wind.

6. Cosmic-Ray Experiments

The Astronomy Survey Committee recommends a series of cosmic-ray experiments in space, to promote the study of solar and stellar activity, the interstellar medium, the origin of the elements, and violent solar and stellar processes.

Cosmic rays—high-energy particles from space that include the nuclei of all elements of the periodic table, as well as electrons and antiparticles—are a sample of contemporary matter from regions far beyond the solar system that we can study in detail. For example, it has now been established that both neon and magnesium in cosmic rays are enriched in their heavier isotopes by comparison with standard solar-system abundances. In their composition and energy spectra, cosmic rays carry unique information about the origin of the elements in stars, the nature of cosmic particle accelerators, and the interstellar gas. Cosmic-ray research has now reached the stage of maturity that permits substantial advances in all of these areas.

At lower energies, and for the lighter elements, some of the required studies could be carried out through the Explorer satellite program, as noted earlier (Section B.1). However, determinations of abundances of the rarer isotopes (such as those of elements heavier than nickel), of the chemical composition of the ultra-heavy nuclei, and of the elemental composition and spectra of lighter elements at high energies, all require large collecting areas and the longest feasible exposure times. Dramatic progress in detector technology during the past decade, together with the ability of the Space Shuttle to place heavy payloads in orbit, now makes it feasible to carry out such measurements with high accuracy over a broad range of energies.

The Committee therefore recommends that NASA begin a systematic program that would pursue the development and construction of the large instruments required to perform these significant cosmic-ray measurements and that would ultimately provide long-duration (approximately 6 months or more) exposures of these instruments in the most appropriate way, possibly through use of a space platform. The development of these large instruments for long-duration exposure will in many cases best be carried out through balloon and Spacelab flights of short duration and through use of ground-based facilities such as the Bevalac accelerator.

7. Astronomical Search for Extraterrestrial Intelligence

The Astronomy Survey Committee recommends an astronomical Search for Extraterrestrial Intelligence (SETI), supported at a modest level, undertaken as a long-term effort rather than as a short-term project, and open to the participation of the general scientific community.

It is hard to imagine a more exciting astronomical discovery or one that would have greater impact on human perceptions than the detection of extraterrestrial intelligence. After reviewing the arguments for and against SETI, the Committee has concluded that the time is ripe for initiating a modest program that might include a survey in the microwave region of the electromagnetic spectrum while maintaining an openness to support of other innovative studies as they are proposed.

Since the chance for a successful detection in the next decade is quite uncertain and may be small, it should be understood that the SETI effort is to be undertaken on a long-term, evolutionary basis. In the coming decade it has been proposed to use a million-channel analyzer in the microwave region of the spectrum; such a program seems appropriate for an initial assay. While exploration of this region of the electromagnetic spectrum appears to make best use of our current technology, other approaches—such as searches using radiation of shorter wavelengths—can also provide interesting opportunities, and the SETI program should allow for the possible support of the ingenious new ideas and proposals that are likely to appear.

Modest support of such programs by U.S. funding agencies is a legitimate scientific activity, and choice of programs within each agency should be made through the normal process of peer review. As a number of other nations have initiated steps toward SETI programs, the opportunities for international collaboration are substantial.

C. SMALL NEW PROGRAMS

A 10-Meter Submillimeter-Wave Antenna

The Astronomy Survey Committee recommends the construction of a submillimeter-wave telescope of about 10-m aperture at a high, dry site.

Recent advances in the design and fabrication of ultraprecise antennas and low-noise receivers make possible an instrument that can observe a substantial portion of the almost completely unexplored submillimeter-wavelength band. Such an instrument will allow detailed study of this rich region of the spectrum with an angular resolution as fine as 8.5 arcsec, which is a factor of 2 better than that of any existing single-element millimeter-wave antenna. The portion of the electromagnetic spectrum accessible to this antenna contains the fine-structure transition of atomic carbon and many rotational lines of important molecules. The facility will be particularly useful for observing higher molecular rotational transitions, such as those of the well-studied interstellar molecules CO and HCN, which are important for cooling processes. Information from a range of molecular levels will make possible a detailed examination of the physical structure of molecular clouds—especially the hot, dynamically active clouds where stars are formed. Since the 10-m telescope will achieve high angular resolution, the transitions of abundant species like carbon and carbon monoxide can be used for studies of objects of small angular extent, such as external galaxies and the envelopes of evolved stars displaying mass loss. The study of the distribution of molecular clouds in external galaxies should, when combined with optical/infrared results, greatly improve our knowledge of star formation in nearby galaxies. If both the 25-Meter Millimeter-Wave Radio Telescope and the 10-m submillimeter-wave antenna are placed at the same site, they may be used together as an interferometer within the short-wavelength range of the former and the long-wavelength range of the latter.

The fascinating problems of star formation, stellar mass loss, and galactic structure stand at the center of many of the controversial issues of astrophysics today. The important information that this instrument can provide on these topics through observations of the more energetic molecular rotational transitions elevates it to an importance much greater than would be suggested by its modest cost.

A Spatial Interferometer for the Mid-Infrared Region

The Astronomy Survey Committee recommends the construction during the early 1980's of a dedicated two-element spatial interferometer for the mid-infrared spectral region with design parameters optimized for operation at a wavelength of 10 μm.

The spectacular success of radio interferometry since World War II has illustrated forcefully the benefits of interferometry to astronomy. Experiments during the last decade have demonstrated the feasibility of spatial interferometry in the infrared region at wavelengths from 2 to 20 μm and have shown that there are a number of important applications for such techniques. Atmospheric propagation characteristics have been shown to be excellent for the 10-μm region in particular, and heterodyne detection techniques have been developed that permit sensitive, coherent detection in this wavelength region.

The Committee supports the construction of a 10-μm heterodyne interferometer for the exciting scientific promise it offers in the near term and as a significant step toward the goal of extending interferometric technology to a wide range of wavelengths. With the later possible addition of an infrared delay line, the interferometer could be made suitable for use with incoherent detectors, when the highest sensitivities to continuum radiation are required.

A dedicated infrared interferometer will have many important astronomical applications. It will permit detailed mapping of infrared sources, many of which are invisible optically because of obscuration by intervening interstellar dust. Of special interest is the circumstellar distribution of dust and molecules shed during episodes of mass loss by many stars during certain stages of their evolution. High-resolution maps of regions of star formation are essential to our understanding of the birth of stars and planetary systems. The angular diameters of large numbers of stars can be measured. Such an interferometer is also well suited to search for an accretion disk around a black hole in the Galactic center—a possibility suggested by recent observations—and to probe the intense infrared sources in the nuclei of active galaxies.

A 10-μm heterodyne interferometer furthermore shows great promise as an astrometric instrument. It will measure precisely the location of infrared sources, aid in the determination of an accurate celestial coordinate system over the entire sky, allow proper motions of stars and other infrared objects to be determined with high precision, and test the General Theory of Relativity to a new order of accuracy by measuring the bending of 10-μm radiation from sources seen near the limb of the Sun.

A Program of High-Precision Optical Astrometry

The Astronomy Survey Committee recommends support for the design and construction of innovative devices that offer the promise of greatly improved astrometric precision, particularly those that may help permit the detection of planets around other stars. Since the 1960's, the typical accuracy of observed stellar parallaxes has improved from ±16 milliarcseconds to ±3 milliarcseconds; the Committee recommends support for the design and construction of innovative devices that offer the potential for obtaining relative positions with an accuracy of ±0.1 milliarcsecond. If the Earth's atmosphere allows the achievement of such accuracy but existing telescopes prove to be inadequate, serious consideration should be given to the construction of a specialized astrometric telescope.

The major advances in astrometry during the past decade have been due to the introduction of new instrumentation and the upgrading of existing instruments. These, together with the advent of new detector technologies and data-analysis techniques, have provided the astronomical community with the tools for decisive advances in the accuracy that can be obtained through ground-based astrometric measurements. For example, one may now look forward to a tenfold improvement in our knowledge of relative stellar positions. Such an advance would permit distance determinations of stars out to 1 kiloparsec, leading to improved luminosities, radii, and masses of stars of virtually all types and ages, including objects such as RR Lyrae stars and members of star clusters that are crucial for calibration of the distance scale to extragalactic objects. Moreover, a substantial improvement in the accuracy of stellar-velocity measurements also seems possible, with important consequences for studies of stellar dynamics.

Such advances can be exploited to search for evidence of extrasolar planetary systems. The detection of such systems would have a significant intellectual impact, removing immediately the apparent uniqueness of our own solar system; systematic observations would enable us to begin the accumulation of data for eventual statistical studies of planetary-formation rates, planetary multiplicity, correlations of physical properties with those of the parent star, and the effects of planetary formation on stellar evolution. An important goal for ground-based astrometry during the coming decade is the detection of Jupiter-like planets around nearby stars, if indeed they exist. The detection of Earth-like planets around other stars appears to require astrometric measurements from space, and the Committee encourages design studies of focal-plane detectors and astrometric telescopes that will yield a positional accuracy sufficient for such measurements.

The Committee also calls attention to the potential for astrometry of a very large space telescope and the advanced interferometers for the radio, infrared, and optical regions of the spectrum recommended for study and development in Chapter 7 of this report.

A Temporary Program to Maintain Scientific Expertise at U.S. Universities

The sharp decline in the anticipated number of university undergraduates in the 1980's, coupled with the unusually small number of faculty retirements anticipated over the same period, will cause a temporary but serious reduction in the number of junior astronomy faculty members that will be appointed by U.S. universities. The intellectual energy that such faculty members bring to astronomy is crucial to progress; moreover, as explained in Chapter 4, basic research at U.S. universities is a critically important component of the national effort in astronomical research. The Astronomy Survey Committee therefore recommends that urgent steps be taken to maintain scientific expertise at U.S. universities by ensuring that excellent younger researchers continue to flow into them during the critical decade ahead.

In particular, the Committee recommends that the Astronomy Division of the National Science Foundation initiate a temporary program of "Astronomy Excellence Awards." Ten to twenty 5-year positions would be awarded to individuals each year on the basis of an open national competition. Each award would be for one half of the salary of a position at the assistant-professor level and would be contingent on commitment of matching funds in the form of the other half of the salary for the same period by a recognized university. It is anticipated that the status and qualifications of successful candidates would be similar to those of regular faculty members at the host institutions and would include the improvements described below. The anticipated cost of this program, which we recommend as a new initiative, would be $0.5 million to $1.0 million per year. It would generate an equal amount of matching funds from universities on a short-term basis and also, the Committee believes, lead to the establishment of new, long-term positions in astronomy. Inasmuch as the problem of declining enrollments and reduced retirements is anticipated to abate starting in about 1990, this program should be terminated at that time.

The Committee furthermore urges the universities themselves to take the following steps to respond to the declining student enrollments and reduced faculty retirements anticipated during the 1980's: implementing procedures that encourage the early retirement of faculty, the establishment of "parallel track" positions of high prestige, and the permitting of non-tenure-track scientists with appropriate qualifications to serve as Principal Investigators on contracts and grants. Some universities have also experimented with so-called "rolling tenure," by which a scientist is granted tenure for the duration of supporting funding from a contract or grant, with tenure extended in step with funding renewal. Since this scheme raises broad issues of university policy, its consideration by a more broadly based committee than the present one would be helpful.

A much more detailed discussion of the issues underlying the present recommendation may be found in the report of the Panel on Organization, Education, and Personnel, to be published in Volume 2 of this survey.

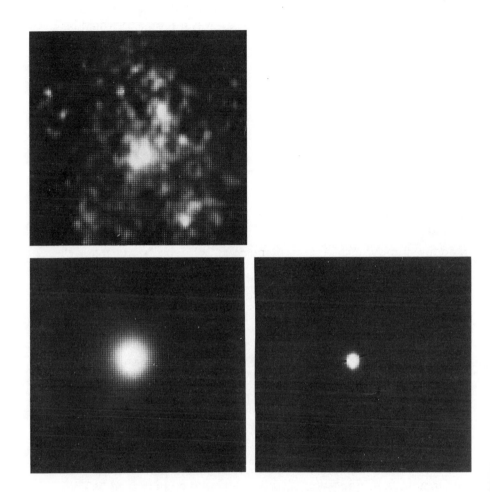

Speckle interferometry permits resolution of a stellar disk. The top frame represents a short-exposure stellar image of ξ Ceti; at the lower right is the speckle image of an unresolved star, ξ Ceti; and at the lower left is the resolved image of the surface of a red giant star, Mira, or o Ceti. (Photo courtesy of R. Stachnik and P. Nisenson, Harvard-Smithsonian Center for Astrophysics)

7

Programs for Study and Development

In addition to the programs recommended in the preceding chapter for approval and funding during the 1980's, the Astronomy Survey Committee considered in detail many other proposed programs of comparable scientific merit. Some of the most promising of these differ from the programs already recommended primarily through being at an earlier stage of technological development or through falling at a later stage in a logically planned observational program. However, planning and development are often time-consuming, especially for large projects. It is therefore important to begin, during the coming decade, study and development of programs that appear to have exceptional promise for the 1990's and beyond. The support of such programs should include funding for design studies and, where appropriate and timely, for the development and testing of instrumentation. Projects and study areas recommended by this Committee in this category include the following, in which the order of listing carries no implication of priority:

A. Future x-ray observatories in space;

B. Instruments for the detection of gravitational waves from astronomical objects;

C. Long-duration spaceflights of infrared telescopes cooled to cryogenic temperatures;

D. A very large telescope in space for ultraviolet, optical, and near-infrared observations;

E. A program of advanced spatial interferometry in the radio, infrared, and optical spectral regions;

F. Advanced gamma-ray experiments; and

G. Astronomical observatories on the Moon.

The present chapter discusses these areas of study and development and illustrates their role in addressing some of the long-range scientific problems that may now be foreseen.

A. FUTURE X-RAY OBSERVATORIES IN SPACE

The Astronomy Survey Committee recommends the study and development of space observatories beyond the Advanced X-Ray Astrophysics Facility (AXAF) for pursuit of a number of important future goals in x-ray astronomy. These investigations will either demand observational capabilities exceeding those of AXAF or will be more appropriate to observatories incorporating specialized, dedicated instrumentation.

The last decade has been a period of extraordinary advances in x-ray astronomy, culminating in the impressive observations returned by the *Einstein* Observatory. These studies will be accelerated and extended to much fainter objects by AXAF, recommended by this Committee as the major new program of highest priority in all of astronomy for the 1980's. However, a number of important future investigations in x-ray astronomy will demand greater sensitivity, spectral resolution, energy coverage, or time resolution than even this powerful facility can provide. Moreover, the rich variety of objects and problems uncovered by the past decade of x-ray observations points to the need for future studies that are too specialized and diverse to be accomplished by a single observatory. Such studies should include at least the following scientific goals:

1. An all-sky survey at x-ray wavelengths comparable in scope with the Palomar Sky Survey, to facilitate the identification and study of objects in various wavelength regions and to provide the data base for detailed studies of the statistics, distribution, and evolution of x-ray sources.

2. Comprehensive measurements of variability, over a wide range of time scales, of the x-ray emissions from faint Galactic and extragalactic sources. This work will be of crucial importance for a detailed understanding of virtually every kind of

compact x-ray source, ranging from nearby flare stars to distant quasars.

3. Low- to moderate-resolution spectroscopy of faint objects, to obtain such data as x-ray red shifts of very distant clusters of galaxies and plasma diagnostics for large numbers of x-ray sources.

4. Detailed study of the large-scale x-ray structure of the Universe through measurements of diffuse features as well as study of structure on smaller scales through observations of clusters of galaxies.

5. High-resolution spectroscopy and polarimetry of faint objects to determine elemental and ionic abundances, temperature distributions, morphologies, and dynamical behaviors. The targets of such studies will include stellar coronas, supernova remnants, active galactic nuclei, and galaxy clusters.

6. Comprehensive investigation of x-ray sources in the energy range above 10 keV, which will be inaccessible to AXAF. Previous balloon and satellite studies have demonstrated the importance of high-energy x-ray measurements in the analysis of the physical conditions in x-ray stars and active galactic nuclei and in the evaluation of the contributions of discrete and diffuse sources to the high-energy x-ray background.

7. Detection, location, and detailed study of transient x-ray sources, including sources of hard x rays associated with gamma-ray bursts and supernova explosions in external galaxies.

8. Measurements of the spectra and variability of soft x-ray and extreme-ultraviolet (EUV) sources, including stellar coronas, single hot white dwarfs, and accreting white dwarfs in binary systems (cataclysmic variables), together with detailed mapping of soft x-ray and EUV emission from the interstellar medium.

The first four of these scientific goals (an all-sky survey, measurement of variability, low- to moderate-resolution spectroscopy, and study of low-surface-brightness features, all directed at faint sources) require collecting areas much larger than that of AXAF but angular resolutions sufficient only to avoid source confusion and to permit unambiguous identification of optical and radio counterparts. One may therefore pursue these goals through deployment of a moderate-resolution array of reflectors of large area, at a cost modest by comparison with that of a monolithic-mirror telescope. Individual modules of such arrays can be designed, tested, and deployed as part of the Spacelab program.

The Committee encourages the continued development of large-area modular arrays of reflectors through Spacelab flights and the eventual assembly of a large-area facility for long-duration observations, perhaps upon a space platform.

High-resolution x-ray spectrometry and polarimetry of faint objects will require specialized instruments with very large effective areas. Preliminary design studies suggest that both kinds of measurements can be pursued in the same long-duration mission, and the Committee urges continued development of the concepts for such a mission.

The study of objects that are most luminous at x-ray energies above 10 keV was begun through balloon and HEAO-1 observations, but a greatly expanded and more systematic study of such sources will be needed during the 1990's. Instruments providing imaging spectrophotometry up to 100-keV photon energies, with concurrent precise location and high-resolution x-ray spectroscopy of such sources, are now becoming technologically feasible; these and other instruments should be studied for possible incorporation into such a mission, which may prove to be an attractive candidate for location on a space platform.

A comprehensive study of the brighter members of the various classes of transient and variable sources will be initiated by the X-Ray Timing Explorer (XTE) satellite. It is already evident, however, that a long-term observatory with enhanced capabilities will be needed during the 1990's to extend the observation of variability to a substantial fraction of the objects accessible to AXAF.

Finally, the Committee believes that planning should begin for soft x-ray and EUV observations beyond those anticipated from the EUV Explorer. Such observations, with improved sensitivity, angular resolution, and spectral resolution, will be needed for analytical studies of the objects detected by EUVE, as well as to complement the optical and near-ultraviolet observations made by Space Telescope and the far-ultraviolet observations to be made by the far-ultraviolet spectrograph in space recommended as a new program in the present report. Coordinated observations of a wide variety of nearby stars at x-ray, EUV, UV, and optical wavelengths will furnish decisive tests of stellar-atmosphere models and theories of coronal activity. Measurement of the structure and physical conditions in the hot component of the interstellar medium by means of soft x-ray and EUV surveys and spectrometry with high angular resolution will be an essential part of future studies of star formation and chemical evolution in the Galaxy.

B. INSTRUMENTS FOR THE DETECTION OF GRAVITATIONAL WAVES

The Astronomy Survey Committee recommends the study and development of instruments for the detection of gravitational waves from astronomical objects.

Detection and measurement of gravitational waves would be of great importance to the achievement of a fundamental understanding of the gravitational field. Success would provide information on important astronomical phenomena that is otherwise unobtainable. This includes, in particular, rapid dynamical behavior in stars and other massive objects and advanced evolution of stellar systems. The great potential value of gravitational-wave detection and the challenging technical problems it presents justify current effort on the study and development of detectors and on the theory of the generation of these waves. When this development reaches a state that allows astronomical observations to become practical, instruments for such observations should be constructed. These will probably include long-baseline systems in space, which at present seem to promise the greatest sensitivity. Meanwhile, advantage should be taken of opportunities for improvements in space-based efforts to detect gravitational waves by upgrading the quality of equipment for tracking the motions of solar-system space probes.

C. LONG-DURATION SPACEFLIGHTS OF INFRARED TELESCOPES COOLED TO CRYOGENIC TEMPERATURES

The Astronomy Survey Committee recommends the study and development of long-duration space observatories incorporating infrared telescopes cooled to cryogenic temperatures. In particular, NASA should study the ways in which the Shuttle Infrared Telescope Facility (SIRTF) may be most effectively incorporated into such a long-duration observatory following a vigorous observational program carried out through Spacelab flights.

During the next decade, the most dramatic gains in astronomical sensitivity will probably occur in the infrared region of the spectrum as a result of observations carried out in space with improved detectors and cryogenically cooled optical systems. Programs currently under development, such as the Infrared Astronomy Satellite (IRAS) and SIRTF, will exploit these advantages in part; by the end of the decade, infrared astronomy should rival radio, optical, and x-ray astronomy in the depth and richness of its observable sky. It is therefore important to begin planning now for the long-duration infrared observatories employing cryogenic optics that will be needed to exploit these opportunities during the 1990's and beyond.

As emphasized in Chapter 4, SIRTF will be the cornerstone of space infrared astronomy during the coming decade. This facility will be able to carry out a powerful observational program even within the relatively brief, 7-day length of an early Shuttle sortie mission. Frequent Shuttle flights of SIRTF, with appropriate refurbishment between flights, thus constitute one of the highest priorities for U.S. astronomy during the 1980's.

However, the scientific return from SIRTF will almost certainly provide an overwhelming case for long-duration spaceflights of cryogenically cooled infrared telescopes. A possible way to achieve this capability is to develop SIRTF itself as a facility for eventual placement onto a space platform, to be refurbished at intervals of 6 to 12 months. Alternatively, development of a free-flying observatory for SIRTF may prove to be more effective. We urge NASA to pursue the general development and design of detectors and facilities for infrared astronomy with the aim of achieving long-duration spaceflights of cryogenic infrared telescopes by 1990.

D. VERY LARGE TELESCOPE IN SPACE FOR ULTRAVIOLET, OPTICAL, AND NEAR-INFRARED OBSERVATIONS

The Astronomy Survey Committee recommends the study and development of the technology required to place a very large telescope in space early in the next century.

The advances in ultraviolet, optical, and infrared astronomy expected in the 1980's are extremely impressive and will certainly lead to great advances in our understanding of diverse astrophysical phenomena. Yet, just as certainly, the instruments of the 1980's will discover new phenomena that will require new and even more powerful facilities for their systematic investigation. By the turn of the century it may be possible to place in orbit a very large telescope—perhaps 30 m in diameter—with diffraction-limited performance from the far-infrared to the near-ultraviolet regions.

The capabilities of such an instrument would be awesome by today's standards. At visual wavelengths it would have a resolving power of about 4 milliarcseconds, a limiting magnitude for medium-resolution spectroscopy of about 30, and a limiting magnitude for imaging of about 33—more than a hundred times fainter than the faintest object accessible to Space Telescope. A telescope of such power could, for example, observe the planets with resolutions ranging from 12 km at Jupiter to 90 km at Neptune, permitting long-term studies of the atmospheric dynamics of the major planets; observe the star-forming regions in the Orion nebula with a resolution of 10 astronomical units at near-infrared wavelengths, permitting direct observation of the process of star formation and detailed studies of preplanetary nebulae; observe the main sequence in the Andromeda Galaxy four magnitudes below the oldest turn-off point and obtain spectra one magnitude below this point; perform spectroscopy of solar-type stars in nearby galaxies, permitting direct chemical-abundance and abundance-history determinations for these galaxies; with suitable care and stability, obtain relative positions of stars and background quasars to a precision of about 100 microarcseconds—enough to provide accurate parallaxes out to a distance of 1000 parsecs and to measure proper motions of stars in globular clusters and in nearby spiral galaxies, permitting studies of the structure and dynamics of these stellar systems; obtain high-resolution ultraviolet spectra of the brightest stars in galaxies of the Virgo cluster of galaxies, making possible detailed studies of the intergalactic medium over long path lengths; and observe the nearest quasars with a resolution of 10 parsecs and the nearest active galaxies with a resolution of 0.2 parsec, fine enough to reveal fundamentally new structural details.

These examples are only a few from a long list of important problems that a very large space telescope could pursue. Such an instrument could be the most powerful tool of astronomy at the start of the next millenium. We therefore recommend that NASA begin exploratory studies of this project and encourage the development of the requisite technologies. Development of technology for the Large Deployable Reflector in space (recommended in Chapter 6) should provide a strong impetus for the further advances needed for this even larger, more powerful instrument.

E. PROGRAM OF ADVANCED SPATIAL INTERFEROMETRY IN THE RADIO, INFRARED, AND OPTICAL SPECTRAL REGIONS

The Astronomy Survey Committee recommends the study and development of advanced spatial interferometers for the radio, infrared, and optical spectral regions.

Exploration of the structural details of astronomical objects from planets to distant quasars requires ever higher angular resolution. Radio astronomers have pioneered in such studies with refined interferometric techniques. The Very Large Array produces radio images with resolutions of an arcsecond or better, and continent-spanning very-long-baseline interferometry currently resolves detail down to a few milliarcseconds. Increases in baselines for radio telescopes and extension of interferometric techniques to infrared and optical wavelengths can achieve even higher resolution.

Important next steps are the Very-Long-Baseline (VLB) Array, the placement of a radio antenna in low-Earth orbit to complement the ground-based VLB Array, and the development of a spatial interferometer for the mid-infrared region, as presented in the recommendations for new programs (Chapter 6). These steps would be followed by the placement of radio antennas in highly elliptical orbits (increasing VLBI baselines by a factor of 10) and by the construction of a ground-based infrared interferometric array with baselines of a few kilometers. These instruments would achieve angular resolutions of a few hundred microarcseconds in the radio and a few milliarcseconds in the infrared region. In both cases, detailed image reconstruction would be possible since they would provide good coverage in the Fourier-transform plane, a necessary condition for the accurate mapping of complex sources.

Further improvements of infrared interferometric capabilities and significant developments in optical interferometry will probably require interferometers in space to escape the deleterious effects of the Earth's atmosphere. A desirable goal is a program of space optical and infrared interferometry leading, by the early part of the next century, to an interferometer with baselines of a few tens of kilometers and resolutions of 1 to 10 microarcseconds at optical wavelengths. Development of interferometric techniques to this extent will require the capability to place large structures in space, significant strides in optical technology, and extensive advances in systems for space-vehicle control, communication, and information processing. With space-based infrared and optical interferometers it would be possible to measure trigonometric parallaxes of objects throughout the Galaxy, to detect Earth-like planets orbiting nearby stars, and to resolve significant structure in nearby quasars, to name only a few of the many important projects that would be advanced with such powerful instruments.

Advanced interferometry on the ground, and then in space, will be among the most important and exciting areas of astronomy in the coming decades. We therefore strongly recommend that NSF and NASA begin preliminary planning, exploratory studies, and technological development for such a program.

F. ADVANCED GAMMA-RAY EXPERIMENTS

The Astronomy Survey Committee recommends the study and development of advanced gamma-ray experiments to follow the program to be carried out by the Gamma Ray Observatory (GRO).

Subsequent to GRO, an advanced high-energy gamma-ray telescope of very large area, high sensitivity, and high angular resolution will be needed for long-term observations of selected sources and regions of special interest. This will be necessary to achieve the statistical accuracy in the counting of gamma-ray photons required to resolve spatial and spectral features of the sources and to analyze their variations. The field of view of the telescope need not be wide, and an appropriate goal for angular resolution is the order of 1 to 2 arcmin. A high-resolution nuclear gamma-ray spectrometer should be included in the mission for the study of the gamma-ray lines from radioactivity in supernova remnants, positron annihilation in the Galactic disk and in extragalactic sources, nuclear excitations caused by cosmic rays in dense matter, and nucleosynthesis in extragalactic supernovae; energy resolution sufficient to study line profiles will be desirable. Development of such instruments, possibly for deployment upon a space platform, should begin as soon as there emerges a clear understanding of the observational requirements from analysis of results from GRO.

G. ASTRONOMICAL OBSERVATORIES ON THE MOON

The Astronomy Survey Committee recommends that agencies of the U.S. Government, working in concert with those of other nations, take steps to ensure the preservation of sites on the Moon (particularly on the far side) for astronomical observations. In addition, the Committee urges NASA to set aside resources for the establishment of lunar astronomical observatories as an important corollary to the establishment of large-scale industrial or power-generation facilities in Earth orbit.

The Moon offers certain decisive advantages as a base for astronomical observations. In particular, the far side of the Moon provides protection from radio interference from sources on or near the Earth and therefore has great potential value for radio astronomy. Shielded at all times from earthlight, sites on the far side of the Moon are also shielded from sunlight for substantial parts of each month and thus offer advantages for optical and infrared observations requiring the darkest possible sky. These considerations become compelling if large military, industrial, or power-generation facilities are constructed in Earth orbit, for the electromagnetic pollution arising from the operation of such facilities may well make it difficult or impossible to observe faint astronomical objects from the ground or from Earth orbit. The utility of the Moon for astronomical observations must therefore be protected as a unique resource for future generations on Earth.

The preservation of sites on the Moon for astronomical observations is clearly a task of international scope. However, the Committee recommends that agencies of the U.S. Government take the lead in such an effort and begin planning in the near future for the establishment of lunar observatories early in the next century. In addition, the U.S. Government should consider carefully the potentially disruptive effects on Earth-based astronomy of the construction of large-scale military, industrial, or power-generation facilities in Earth orbit.

Appendixes

Appendix A

Statement Concerning a Space Platform

Much of modern astronomy must be done above the Earth's atmosphere. Because the resources demanded by such investigations are substantial, the Committee therefore devoted considerable discussion to ways in which space science might be carried out with greater flexibility and at lower cost. The currently developing concept of a "space platform" offers considerable promise in these respects.

In addition to important aircraft and balloon facilities for observations above most of the Earth's atmosphere, there will be in operation by the early 1980's three quite different types of space-science vehicles, each providing observations on a different time scale. Sounding rockets, the first of these to be developed, will still be important for space exposures requiring only a few minutes' duration. They offer great flexibility in location, launch timing, and payload content, also providing valuable opportunities for developing satellite instrumentation and for training space scientists at low cost. The Space Shuttle will powerfully augment U.S. space-astronomy capability by offering orbital exposures on Spacelab ranging effectively from hours to a few days, and it will also accommodate large payloads; however, among the larger experiments, only a few (such as SIRTF and SOT) can carry out their missions with maximum cost-effectiveness within such relatively brief exposure times. Free-flying satellites often present the most advantageous means for carrying out major scientific programs, permitting dedicated, noninterfering payloads and observing lifetimes ranging up to years. However, each

169

individual spacecraft is expensive and (except for major observatories such as ST and AXAF) not normally accessible for refurbishment, modification, or recovery after launch.

In terms of exposure times provided, there is a large gap in capability between Spacelab missions and those carried out on free-flying satellites. On the other hand, many areas of space astronomy require long-duration, relatively low-cost exposure, together with large payload capacity and accessibility for replenishment of expendable materials, for repair or replacement of components, and for return to Earth for reconfiguration.

A space platform could in principle supply these needs. Current concepts envisage free-flying structures, designed for lifetimes of at least a decade, consisting of a central module (containing control-moment gyroscopes and communications equipment) attached to a substantial solar-power unit. Appropriate docking fixtures would permit the simple attachment of numbers of pallets functionally similar to, if not identical to, those used for Spacelab experiments. Platform extension arms could be used to reduce interference between experiments, most of which would carry their own pointing systems for the requisite precision. Shuttle flights, perhaps including some scheduled for other purposes, would be able to visit the platform several times a year to reprovision expendable materials and to repair or replace experimental hardware.

A space platform appears to offer many advantages to other scientific areas as well, such as biomedical research and materials processing. Regardless of the influence of such other fields on possible platform design, the Astronomy Survey Committee urges that at least one line of space-platform evolution be guided strongly by the needs of observational astronomy. Specifically, platforms optimized for astronomical use must offer unmanned operation, simplicity and economy of both construction and operation, and the convenient servicing and replacement of experiments. The Committee recommends that NASA continue to seek the advice and recommendations of the scientific community throughout the development of a space platform for observational astronomy.

Several areas of astronomy could profit substantially from platforms dedicated entirely to their use. For example, a cluster of solar experiments could fill a solar-oriented platform in polar orbit. Another set of experiments, with capability for imaging, spectroscopy, and studies of time variation in the gamma-ray, x-ray, ultraviolet, and infrared regions, would be well suited to a platform observatory. Some types of astronomical research, such as cosmic-ray studies, would be expected to place few constraints on platform characteristics and the choice of neighboring experiments.

Various scientific study groups have already identified many astronomical missions that would appear to be substantially more cost-effective if flown on a space platform, rather than on Spacelab or a free-flying satellite. The Committee commends NASA's initiative in studying the platform concept and emphasizes that these studies should include, for astronomy, consideration of the simplest and least expensive system able to carry out basic platform functions.

Appendix B _____

Organization, Education, and Personnel

During its study of the needs of astronomy and astrophysics for the 1980's, the Astronomy Survey Committee devoted considerable attention to the general structure and health of the profession, particularly through discussion of the recommendations of the Panel on Organization, Education, and Personnel (OEP) presented in Volume 2.

An investigation of the trends that training and employment patterns in astronomy will follow in the future led the Panel to its first and most important recommendation, one that has also been adopted by the Committee itself as a recommended new program for the coming decade: a temporary program to maintain scientific expertise at U.S. universities through a series of NSF "Astronomy Excellence Awards" during the 1980's. A discussion of this recommendation appears in Section C of Chapter 6.

The Astronomy Survey Committee also supports the other recommendations of the OEP Panel, which are discussed extensively in the OEP Panel report and summarized below.

PERSONNEL

1. *Minorities* The Panel endorses the recommendations made by the American Astronomical Society's Committee on Ethnic Minorities to encourage young members of ethnic minorities to study astronomy. Past progress in this area has been inadequate.

172

2. *Women in Astronomy* The Panel endorses the report of the Committee on the Status of Women, accepted by the American Astronomical Society in 1980. Women are still far from achieving equal status in astronomy.

3. *Dual-Career Couples* The Panel recommends appropriate modification of remaining nepotism rules, the granting of permission to scientists employed part-time to act as Principal Investigators on contracts and grants, and the liberalization of institutional policies governing shared jobs.

EDUCATION

4. *Public Communication* The Panel recognizes the need for astronomers to devote a suitable portion of their time to the communication of astronomical results to the general public, and encourages them to do so. Such efforts need to be recognized and encouraged also by department chairs and group leaders, funding agencies, academic institutions, and professional organizations as a necessary and beneficial scientific service activity.

5. *Training of Astronomers* The Panel recommends that the training of astronomers include the acquisition of skills in such specialized areas as electronics, electrooptical devices, mechanical systems, computer software, and systems engineering; these skills not only are relevant to the development of astronomical instrumentation but also make astronomy graduates more attractive to industry. There is a perception that astronomers who develop advanced astronomical instrumentation are sometimes not adequately rewarded with respect to promotion and tenure. The Panel recommends that astronomy departments take care to eliminate any such inequity.

6. *The Astronomical Community* Teachers at two- and four-year colleges, many of whom are not full-time astronomers, make a significant contribution to astronomy education in the United States. Amateur astronomers also contribute in substantial ways to the position astronomy holds in the national esteem. The Panel recommends that research astronomers make efforts to increase communication with these additional members of the astronomical community, who contribute so much to the general health of the field.

7. *Small Telescopes* Small telescopes, many associated with university departments, are an important resource for U.S. astronomy. Financial support for these telescopes (and associated instrumentation) should be awarded on the basis of scientific

merit. In awarding funds, agencies should keep in mind the many diverse needs served by these facilities.

ORGANIZATION

8. *Classified Data and Technology* The Panel recommends that both NASA and NSF maintain a continuing awareness of the benefits that would accrue to astronomy from the use of certain data and technology that have been classified, inform the proper government agencies of such benefits, and establish appropriate mechanisms by which the astronomical community can participate in the procedures for identification and declassification of such data and technology.

9. *Access to Foreign Space Missions on the Basis of Merit* The Panel recommends that NASA work to promote competitive access to foreign scientific satellite missions and institute policies and budgetary mechanisms designed to encourage the flight of U.S. experiments on foreign satellites.

10. *Peer Review* The Panel calls attention to the study by J.R. Cole, L.C. Rubin, and S. Cole (*Scientific American*, October 1977, p. 34), which "yielded little evidence in support of the main criticisms that have been made of the peer-review system." The Panel supports any measures that can be taken to streamline proposal procedures but recognizes that increased accountability requirements are beyond the direct control of the astronomical community (e.g., E.B. Staats, *Science 205*, 18, July 6, 1979). The Panel also emphasizes the great importance of supporting projects whose results may lie far in the future and the particular need for dialogue between the proposer and referees when instrumental proposals are under review. Finally, the Panel notes the importance of attracting outstanding scientists to work within the federal funding agencies and of opportunities for temporary agency service under the Intergovernmental Personnel Act. The welfare of the entire astronomical community depends critically on the wisdom and foresight of scientific decisions made within federal agencies.

11. *Advice to NASA and NSF* The Panel recommends that the agency that funds a scientific mission should take particular care also to fund adequate analysis of all the meaningful data that flow from that mission. Using the mechanisms for interagency cooperation already in place, the agencies should identify the mutual impact of new programs before those programs are initiated and take appropriate action.

12. *Private and State Support for Astronomy* The Panel commends the successful efforts of institutions that have done well in this area. A number of state universities have been notably successful in obtaining funds specifically designated for astronomy from their state legislatures. Private institutions have also provided substantial support for astronomy; a number have been particularly successful in maintaining strong research programs in spite of the inroads of inflation.

13. *Reduced Administrative Burdens and Multiyear Funding* The Panel urges funding agencies to switch, as rapidly as possible, to longer-term (e.g., three-year) funding of research projects, with reporting requirements reduced to submission of copies of published papers, annual reports, or both. The Panel further urges that simple mechanisms be instituted for consolidation of small projects from a single agency.

Appendix C

Panels and Working Groups

PANEL ON HIGH ENERGY ASTROPHYSICS

GEORGE W. CLARK, Massachusetts Institute of Technology, *Chairman*
C. STUART BOWYER, University of California, Berkeley
RICCARDO GIACCONI, Space Telescope Science Institute
ALLAN S. JACOBSON, Jet Propulsion Laboratory
WILLIAM L. KRAUSHAAR, University of Wisconsin
DIETRICH MUELLER, University of Chicago
REUVEN RAMATY, NASA Goddard Space Flight Center
DAVID SCHRAMM, University of Chicago
KIP THORNE, California Institute of Technology

CARL E. FICHTEL, NASA Goddard Space Flight Center, *ex officio*
ARTHUR B. C. WALKER, Stanford University, *ex officio*

PANEL ON ULTRAVIOLET, OPTICAL, AND INFRARED ASTRONOMY

E. JOSEPH WAMPLER, University of California, Santa Cruz, *Chairman*
JACQUES BECKERS, University of Arizona
GEOFFREY BURBIDGE, Kitt Peak National Observatory
GEORGE CARRUTHERS, U.S. Naval Research Laboratory
JUDITH G. COHEN, California Institute of Technology

176

JOHN GALLAGHER, University of Illinois, Urbana
FRED GILLETT, Kitt Peak National Observatory
W.A. HILTNER, University of Michigan
WILLIAM F. HOFFMANN, University of Arizona
JEFFREY LINSKY, Joint Institute for Laboratory Astrophysics and the University of Colorado
J. BEVERLEY OKE, California Institute of Technology
VERA RUBIN, Carnegie Institution of Washington
RAINER WEISS, Massachusetts Institute of Technology
SIDNEY C. WOLFF, University of Hawaii
DONALD YORK, Princeton University

Consultants

J. ROGER ANGEL, University of Arizona
JESSE GREENSTEIN, California Institute of Technology
LYMAN SPITZER, Princeton University
STEPHEN E. STROM, Kitt Peak National Observatory

PANEL ON RADIO ASTRONOMY

PATRICK THADDEUS, NASA Goddard Institute for Space Studies and Columbia University, *Chairman*
BERNARD BURKE, Massachusetts Institute of Technology
MARSHALL COHEN, California Institute of Technology
FRANK DRAKE, Cornell University
MORTON ROBERTS, National Radio Astronomy Observatory
JOSEPH TAYLOR, Princeton University
WILLIAM J. WELCH, University of California, Berkeley
DAVID WILKINSON, Princeton University
ROBERT WILSON, Bell Laboratories

Consultant

GEORGE A. DULK, University of Colorado

PANEL ON THEORETICAL AND LABORATORY ASTROPHYSICS

RICHARD A. McCRAY, Joint Institute for Laboratory Astrophysics and the University of Colorado, *Chairman*
W. DAVID ARNETT, University of Chicago

ROGER BLANDFORD, California Institute of Technology
ALEXANDER DALGARNO, Harvard-Smithsonian Center for Astrophysics
WILLIAM FOWLER, California Institute of Technology
WILLIAM PRESS, Harvard-Smithsonian Center for Astrophysics
SCOTT D. TREMAINE, Massachusetts Institute of Technology
JAMES G. WILLIAMS, Jet Propulsion Laboratory

Consultants

ARTHUR N. COX, Los Alamos Scientific Laboratory
KRIS DAVIDSON, University of Minnesota
VICTOR G. SZEBEHELY, University of Texas, Austin
C. BRUCE TARTER, Lawrence Livermore Laboratory

PANEL ON DATA PROCESSING AND COMPUTATIONAL
FACILITIES

EDWARD J. GROTH, Princeton University, *Chairman*
ROBERT M. HJELLMING, National Radio Astronomy Observatory
RICHARD B. LARSON, Yale University
JAYLEE M. MEAD, NASA Goddard Space Flight Center
RICHARD H. MILLER, University of Chicago
BERNARD OLIVER, Hewlett-Packard Corporation
STEPHEN E. STROM, Kitt Peak National Observatory
PAUL R. WOODWARD, Lawrence Livermore Laboratory

PANEL ON ORGANIZATION, EDUCATION, AND
PERSONNEL

RICHARD C. HENRY, The Johns Hopkins University, *Chairman*
PETER B. BOYCE, American Astronomical Society
NOEL W. HINNERS, Smithsonian Institution
HENRY L. SHIPMAN, University of Delaware
ELSKE V.P. SMITH, Virginia Commonwealth University
DONNA E. WEISTROP, NASA Goddard Space Flight Center

Consultants

DONALD W. GOLDSMITH, Interstellar Media
MARTHA H. LILLER, Harvard-Smithsonian Center for Astrophysics

WAYNE OSBORN, Central Michigan University
R. MARCUS PRICE, University of New Mexico, Albuquerque

WORKING GROUP ON SOLAR PHYSICS

ARTHUR B.C. WALKER, Stanford University, *Chairman*
JOHN W. HARVEY, Kitt Peak National Observatory
THOMAS E. HOLZER, National Center for Atmospheric Research
JEFFREY L. LINSKY, Joint Institute for Laboratory Astrophysics and the University of Colorado
EUGENE N. PARKER, University of Chicago
ROGER K. ULRICH, University of California, Los Angeles
GERARD VAN HOVEN, University of California, Irvine
GEORGE L. WITHBROE, Harvard-Smithsonian Center for Astrophysics

Consultants

HUGH S. HUDSON, University of California, San Diego
STUART D. JORDAN, NASA Goddard Space Flight Center
MUKUL R. KUNDU, University of Maryland
JACK B. ZIRKER, Sacramento Peak Observatory

WORKING GROUP ON PLANETARY SCIENCE

MICHAEL J.S. BELTON, Kitt Peak National Observatory, *Chairman*
JOHN J. CALDWELL, State University of New York, Stony Brook
DONALD M. HUNTEN, University of Arizona
TORRENCE V. JOHNSON, Jet Propulsion Laboratory
DAVID MORRISON, University of Hawaii
TOBIAS C. OWEN, State University of New York, Stony Brook
STANTON J. PEALE, University of California, Santa Barbara
GORDON H. PETTENGILL, Massachusetts Institute of Technology
JAMES B. POLLACK, NASA Ames Research Center

WORKING GROUP ON GALACTIC ASTRONOMY

ROBERT D. GEHRZ, University of Wyoming, *Chairman*
DAVID BLACK, NASA Ames Research Center
W. BUTLER BURTON, University of Minnesota
DUANE F. CARBON, Kitt Peak National Observatory
JUDITH G. COHEN, California Institute of Technology

PIERRE DEMARQUE, Yale University
FREDERICK K. LAMB, University of Illinois, Urbana
BRUCE MARGON, University of Washington, Seattle
PHILIP SOLOMON, State University of New York, Stony Brook
SIDNEY VAN DEN BERGH, Dominion Astrophysical Observatory
PETER O. VANDERVOORT, University of Chicago

Consultants

RICHARD A. McCRAY, Joint Institute for Laboratory Astrophysics
and the University of Colorado
CHRISTOPHER F. McKEE, University of California, Berkeley
LEONARD SEARLE, Carnegie Institution of Washington

WORKING GROUP ON EXTRAGALACTIC ASTRONOMY

S. M. FABER, University of California, Santa Cruz, *Chairman*
CHRISTOPHER F. McKEE, University of California, Berkeley
FRAZER OWEN, National Radio Astronomy Observatory
P. JAMES E. PEEBLES, Princeton University
JOSEPH SILK, University of California, Berkeley
HARVEY TANANBAUM, Harvard-Smithsonian Center for Astro-
physics
ALAR TOOMRE, Massachusetts Institute of Technology
JAMES W. TRURAN, University of Illinois, Urbana
RAY J. WEYMANN, University of Arizona

JAMES E. GUNN, Princeton University, *ex officio*
JEREMIAH OSTRIKER, Princeton University, *ex officio*

Consultant

BEATRICE M. TINSLEY, Yale University

WORKING GROUP ON RELATED AREAS OF SCIENCE

JAMES E. GUNN, Princeton University, *Chairman*
DOUGLAS EARDLEY, Harvard-Smithsonian Center for Astro-
physics
PETER GILMAN, National Center for Atmospheric Research
RUSSELL M. KULSRUD, Princeton University
DAVID PINES, University of Illinois, Urbana

GERALD J. WASSERBURG, California Institute of Technology
WILLIAM D. WATSON, University of Illinois, Urbana
STEVEN WEINBERG, Harvard University
STAN E. WOOSLEY, University of California, Santa Cruz

WORKING GROUP ON ASTROMETRY

GART WESTERHOUT, U.S. Naval Observatory, *Chairman*
HEINRICH K. EICHHORN, University of Florida, Gainesville
GEORGE D. GATEWOOD, Allegheny Observatory
JAMES HUGHES, U.S. Naval Observatory
WILLIAM H. JEFFERYS, University of Texas, Austin
IVAN R. KING, University of California, Berkeley
WILLIAM F. VAN ALTENA, Yale University

WORKING GROUP ON THE SEARCH FOR EXTRATERRESTRIAL INTELLIGENCE

HARLAN J. SMITH, University of Texas, Austin, *Chairman*
FRANK DRAKE, Cornell University
JAMES E. GUNN, Princeton University
DAVID HEESCHEN, National Radio Astronomy Observatory
NOEL W. HINNERS, Smithsonian Institution
JEREMIAH OSTRIKER, Princeton University
PATRICK THADDEUS, NASA Goddard Institute for Space Studies
and Columbia University
CHARLES H. TOWNES, University of California, Berkeley
BENJAMIN M. ZUCKERMAN, University of Maryland

Consultants

GEORGE D. GATEWOOD, Allegheny Observatory
MICHAEL HART, Trinity University, San Antonio
MICHAEL D. PAPAGIANNIS, Boston University

Appendix D

Abbreviations Used in Text

AAS	American Astronomical Society
ASO	Advanced Solar Observatory (in space)
AXAF	Advanced X-Ray Astrophysics Facility
CCD	Charge-coupled device
CES	Committee on Earth Sciences (SSB)
COBE	Cosmic Background Explorer satellite
COMPLEX	Committee on Planetary and Lunar Exploration (SSB)
CSAA	Committee on Space Astronomy and Astrophysics (SSB)
CSSP	Committee on Solar and Space Physics (SSB)
EOP	Experiment of Opportunity Program (NASA)
EUVE	Extreme Ultraviolet Explorer satellite
GRIST	Grazing Incidence Solar Telescope
GRO	Gamma Ray Observatory
HEAO	High Energy Astronomical Observatory
IRAS	Infrared Astronomy Satellite (Explorer)
ISPM	International Solar Polar Mission
IUE	International Ultraviolet Explorer satellite
KAO	Kuiper Airborne Observatory
LDR	Large Deployable Reflector (in space; infrared/submillimeter)
MMT	Multiple-mirror telescope (optical/infrared)
NAS	National Academy of Sciences
NASA	National Aeronautics and Space Administration
NBS	National Bureau of Standards

182

NRC	National Research Council
NSF	National Science Foundation
NTT	New Technology Telescope (optical/infrared, ground-based)
OEP	Organization, Education, and Personnel (Panel)
OSSA	Office of Space Science and Applications (NASA)
PI	Principal Investigator
SETI	Search for Extraterrestrial Intelligence
SIRTF	Shuttle Infrared Telescope Facility
SOT	Solar Optical Telescope (Space Shuttle facility)
SSB	Space Science Board
SSXTF	Solar Soft X-Ray Telescope Facility
ST	Space Telescope (optical/ultraviolet)
STSCI	Space Telescope Science Institute
VLA	Very Large Array (radio telescope)
VLB	Very Long Baseline
VLB Array	Very-Long-Baseline Array (of radio telescopes)
VLBI	Very-long-baseline interferometry
XTE	X-Ray Timing Explorer satellite

Index